Challenges in Foundation Engineering - Case Studies and Best Practices

Edited by Mohamed Ayeldeen

Published in London, United Kingdom

Challenges in Foundation Engineering - Case Studies and Best Practices
http://dx.doi.org/10.5772/intechopen.107733
Edited by Mohamed Ayeldeen

Contributors
Ali Asghar Firoozi, Ali Akbar Firoozi, Bhavanasi Subbaratnam, Erol Guler, Mohamed Ayeldeen, P. Francisco Centeno

First published in London, United Kingdom, 2024 by IntechOpen
IntechOpen is the global imprint of INTECHOPEN LIMITED, registered in England and Wales, registration number: 11086078, 167-169 Great Portland Street, London, W1W 5PF, United Kingdom

British Library Cataloguing-in-Publication Data
A catalogue record for this book is available from the British Library

Additional hard and PDF copies can be obtained from orders@intechopen.com

Challenges in Foundation Engineering - Case Studies and Best Practices
Edited by Mohamed Ayeldeen
p. cm.

This title is part of the Civil Engineering Book Series, Volume 8
Topic: Structural Engineering
Series Editor: Assed Haddad
Topic Editor: Wael A. Zatar

Print ISBN 978-0-85014-210-5
Online ISBN 978-0-85014-211-2
eBook (PDF) ISBN 978-0-85014-212-9
ISSN 3029-0287

For EU product safety concerns:
IN TECH d.o.o., Prolaz Marije Krucifikse Kozulić 3, 51000 Rijeka, Croatia,
info@intechopen.com or visit our website at intechopen.com.

IntechOpen Book Series

Civil Engineering

Volume 8

Aims and Scope of the Series

Civil engineering is a traditional field of engineering from which most other branches of engineering have evolved. It comprises traditional sub-areas like transportation, structures, construction, geotechnics, water resources, and building materials. It also encompasses sustainability, risk, environment, and other concepts at its core. Historically, developments in civil engineering included traditional aspects of architecture and urban planning as well as practical applications from the construction industry. Most recently, many elements evolved from other fields of knowledge and topics like simulation, optimization, and decision science have been researched and applied to increase and evolve concepts and applications in this field. Civil engineering has evolved in the last years due to the demands of society in terms of the quality of its products, modern applications, official requirements, and cost and schedule restrictions. This series addresses real-life problems and applications of civil engineering and presents recent, cutting-edge research as well as traditional knowledge along with real-world examples of developments in the field.

Meet the Series Editor

Professor Assed N. Haddad is a Civil Engineer with a degree from the Federal University of Rio de Janeiro (UFRJ) earned in 1986, as well as a Juris Doctor degree from the Fluminense University Center earned in 1993, and a Master's degree in Civil Engineering from the Fluminense Federal University (UFF) obtained in 1992. He completed his Ph.D. in Production Engineering from COPPE / Federal University of Rio de Janeiro in 1996. Professor Haddad's academic pursuits have taken him to postdoctoral stays at the University of Florida, USA in 2006; at the Universitat Politècnica de Catalunya, Spain in 2010; and at the University of New South Wales Sydney, Australia in 2019. Currently, he serves as a Full Professor at the Federal University of Rio de Janeiro. He has held visiting professorships at various institutions including the University of Florida, Universitat Politècnica de Catalunya, Universitat Rovira i Virgili, and Western Sydney University. His research expertise encompasses Civil, Environmental, and Production Engineering, with a primary focus on the following topics: Construction Engineering and Management, Risk Management, and Life Cycle Assessment. He has been the recipient of research grants from the State of Rio de Janeiro, Brazil: CNE FAPERJ from 2019 to 2022 and from 2023 to 2025. Additionally, his research grants obtained from the Brazilian Government CNP since 2012 last to this date. Professor Haddad has been involved in several academic endeavors, being the Guest Editor of the International Journal of Construction Management; MDPI's Sustainability, Energies, and Infrastructures; Associate Editor at Frontiers in Built Environment / Sustainable Design and Construction; Guest Editor at Frontiers in Built Environment / Construction Management; and Academic Editor of the Journal of Engineering, Civil Engineering Section of Hindawi. He is currently a Professor of the Environmental Engineering Program at UFRJ and the Civil Engineering Program at UFF.

Meet the Volume Editor

Dr. Mohamed Ayeldeen boasts more than 18 years of extensive experience in geotechnical engineering, coupled with a profound understanding of geoenvironmental engineering. With a notable portfolio of iconic infrastructure projects under his belt, he has left an indelible mark on the field. Dr. Ayeldeen's expertise spans across various regions worldwide, where he has adeptly managed and directed large-scale geotechnical endeavors. His wealth of experience encompasses the design and execution of complex projects, reflecting his versatility and proficiency in navigating diverse challenges. Moreover, Dr. Ayeldeen has made significant contributions to the academic community, with numerous papers published in esteemed international journals and conferences. His scholarly works have garnered widespread recognition, evidenced by more than 600 citations, further solidifying his stature as a thought leader in the field of geotechnical engineering.

Contents

Preface

Welcome to Challenges in Foundation Engineering – Case Studies and Best Practices. Within these pages, you will find a curated collection of chapters that delve deep into the multifaceted realm of foundation engineering. Each chapter offers invaluable insights, drawing from real-world case studies and best practices to tackle the contemporary challenges facing the field.

This book covers a diverse array of topics, including shallow and deep foundations, soil–structure interaction, and reinforced soil. It serves not merely as a compilation of case studies but also as a comprehensive synthesis of the current state of research in foundation engineering. The interdisciplinary nature of the topics reflects the collaborative spirit necessary to navigate the complexities inherent in the field.

In the first chapter following the introduction of our book "Dynamic Stability of Beam on Elastic Foundation Including Higher Transition Foundation", we explore the dynamic stability of beams on elastic foundation. Employing the energy method, a precise analytical solution has been crafted to anticipate the dynamic instability bounds of simply supported beams on elastic foundations. This analysis zeroes in on the influence of higher transition foundations on dynamic stability boundaries. Findings indicate that as the elastic foundation parameter value increases, the width of dynamically unstable zones decreases, thereby reducing the susceptibility of the beam to dynamic stability phenomena under periodic loads. Additionally, the study demonstrates the existence of master dynamic instability curves by utilizing precise non-dimensional parameters for the imposed periodic load and its radian frequency.

In Chapter 3, "Integrity Assessment for Drill Shafts Foundation in Public and Private Works with Available Technologies in the Twenty-First Century", we focus on integrity control tests to detect anomalies and construction defects promptly. These tests provide precise information on pile geometry using methods like pile impedance, ultrasonic logging, and thermal profiling. They facilitate the identification of anomalies in 2D and 3D, allowing evaluation of concrete placement integrity during hardening. Despite their benefits, these tests also have limitations, which we discuss based on field experience and research findings from global experts and the Deep Foundations Institute.

In Chapter 4, "Soil-Structure Interaction: Understanding and Mitigating Challenges", we delve into the intricate dynamics of soil–structure interaction (SSI). The chapter wraps up by delving into innovative mitigation strategies for challenges associated with SSI, including the utilization of novel materials and advanced computational models. Emphasizing sustainability and resilience, the discussion centers on practices designed to endure the impacts of time and climate change. This comprehensive overview of SSI aims to equip readers with the insights and tools necessary to understand and address the complexities of this crucial aspect of geotechnical engineering. Serving as a valuable reference, the chapter caters to a diverse audience of researchers, engineers, students, and practitioners, facilitating the creation of safer, more resilient structures while minimizing environmental repercussions.

Finally, Chapter 5, "Geosynthetic Reinforcement Applications", serves to introduce readers to various applications of reinforcement. The chapter provides a brief overview of geosynthetics before delving into specific examples such as geosynthetics-reinforced soil retaining walls, which offer benefits under earthquake loading conditions. Additionally, the chapter discusses using geosynthetic reinforcement below foundations to enhance bearing capacity and explores the use of geosynthetics as encasements around stone columns, showcasing improvements in seismic performance compared to traditional stone columns. Through these examples, readers gain insight into the versatility and effectiveness of geosynthetic reinforcement across different engineering contexts.

As you embark on this intellectual journey, I encourage you to adopt a discerning perspective, embracing the dynamic nature of foundation engineering. This book represents more than just a collection of chapters; it stands as a testament to the collective pursuit of knowledge and excellence in geotechnical engineering.

May the wealth of knowledge and insights gleaned from these pages catalyze the development of groundbreaking solutions in foundation engineering. As we navigate the complexities of this field, let us remain committed to fostering sustainable practices and embracing innovation at every turn. By continually pushing the boundaries of our understanding and leveraging the latest advancements in technology, we can forge a path toward a future where foundation engineering not only meets the demands of today but also anticipates and addresses the challenges of tomorrow. Together, let us strive to shape a more resilient and sustainable built environment for generations to come.

Dr. Mohamed Ayeldeen
Technical Director,
Arcadis Consulting,
London, United Kingdom

Chapter 1

Introductory Chapter: Challenges in Foundation Engineering – Case Studies and Best Practices

Mohamed Ayeldeen

1. Introduction

In the ever-evolving landscape of civil engineering, foundation engineers find themselves grappling with an array of challenges that have assumed new dimensions in the twenty-first century. While the global economy traverses through periods of uncertainty, the domain of foundation engineering is not insulated from its far-reaching impacts. Economic challenges have given rise to an intensified focus on efficiency, cost-effectiveness, and the imperative to develop infrastructure that stands resilient in the face of adversities. Consequently, the demand for smart and sustainable solutions has reached unprecedented levels, challenging engineers to innovate and adapt.

Foundations come in a variety of forms, each tailored to suit the specific requirements of the structure and the characteristics of the underlying soil or rock strata. Broadly categorized, foundations can be classified into shallow foundations and deep foundations. Shallow foundations, as the name suggests, are those that typically penetrate only a few meters into the ground and spread their loads over a larger area. These include footings, raft foundations, and mat foundations, among others. Deep foundations, on the other hand, are designed to transfer structural loads to deeper, more competent soil or rock layers. These may include piles, drilled shafts, and caissons, depending on the site conditions and engineering considerations [1, 2].

Regardless of their type, the fundamental function of foundations remains consistent: to distribute the loads from the structure safely into the ground, ensuring stability and preventing settlement or failure. Whether supporting towering skyscrapers in urban landscapes or resilient infrastructure in remote regions, foundations play a pivotal role in the built environment, often operating silently beneath our feet but serving as the critical backbone of our civilization.

2. Geotechnical engineering and its applications

Geotechnical engineering is the cornerstone of foundation design and construction, focusing on the behavior of earth materials—soil and rock—and their interaction with civil engineering structures. Understanding the properties of soil and rock formations is paramount in ensuring the stability, safety, and longevity of foundations. Soil types vary widely, ranging from cohesive clays to granular sands and everything in between. Each type possesses distinct characteristics that influence its behavior under load, moisture content, and other environmental factors. Cohesive soils, such as clay, tend to exhibit high plasticity and low permeability, making them susceptible to swelling, shrinkage, and consolidation. Granular soils, like sand and gravel, offer better drainage but may experience settlement and instability if not properly compacted or reinforced [3].

During foundation construction and design, various soil-related challenges can arise, presenting engineers with complex problems to overcome. One common issue is soil settlement, which occurs when the soil beneath a foundation compress under the weight of the structure, leading to uneven settling and potential structural damage. This phenomenon is particularly prevalent in areas with expansive clay soils or poorly compacted fill materials. Another challenge is soil liquefaction, a phenomenon in which saturated granular soils lose their strength and stiffness during seismic events, behaving more like a liquid than a solid. Liquefaction can cause catastrophic failure of foundations, especially in regions prone to earthquakes, posing significant risks to structures and occupants alike. Additionally, soil erosion and instability can pose challenges during foundation construction, especially in areas with steep slopes, high groundwater tables, or heavy rainfall. Erosion control measures and slope stabilization techniques are essential to mitigate these risks and ensure the long-term stability of foundations [4].

Geotechnical engineers employ a variety of techniques and methodologies to address these challenges, including soil testing and analysis, ground improvement methods, and innovative foundation design approaches. By understanding the unique properties and behavior of soil and rock formations, engineers can develop effective solutions to safeguard structures against the unpredictable forces of nature and ensure the resilience and sustainability of our built environment [5–7].

3. Geotechnical engineering and climate change

As we confront the challenges of climate change and global warming, the role of geotechnical engineering in sustainable development has never been more critical. The construction and infrastructure sectors are significant contributors to greenhouse gas emissions, primarily through energy-intensive processes and the extraction of raw materials. Geotechnical engineering, with its focus on the built environment's interaction with the natural landscape, plays a pivotal role in mitigating these impacts and promoting environmentally responsible practices. One of the key considerations in addressing global warming within the geotechnical industry is the reduction of carbon emissions associated with construction activities. This can be achieved through the adoption of green engineering principles, which prioritize energy efficiency, resource conservation, and environmental stewardship throughout the project lifecycle. By optimizing construction methods, materials selection, and site management practices, engineers can minimize the carbon footprint of foundation projects while maintaining structural integrity and safety [8, 9].

The choice of foundation systems and construction techniques can significantly impact a project's sustainability. Sustainable foundation solutions aim to minimize environmental disruption, conserve natural resources, and enhance the resilience of built structures to climate-related hazards. Innovative approaches such as recycled materials for backfilling, prefabricated foundation components, and bio-based soil stabilization techniques offer sustainable alternatives to traditional construction methods, reducing both environmental impact and project costs [9, 10].

4. Research and development importance

Research and development (R&D) serve as the cornerstone of innovation and progress in all fields, including geotechnical engineering. In the realm of foundation engineering, the pursuit of new knowledge and the application of cutting-edge technologies are essential for addressing the complex challenges facing the industry. However, for R&D efforts to have a meaningful impact, it is crucial to bridge the gap between academic research and practical applications, ensuring that innovative solutions are not only technically sound but also economically viable, time-saving, and environmentally sustainable [11].

However, the translation of academic research into practical solutions faces several challenges, including the complexity of real-world engineering problems, the limitations of existing technologies, and the constraints of time and budget. Bridging this gap requires collaboration and knowledge exchange between academia and industry, fostering a symbiotic relationship that leverages the strengths of both sectors. Practical solutions derived from academic research offer several advantages over conventional approaches. First and foremost, they are often more efficient and cost-effective, leveraging innovative technologies and methodologies to streamline construction processes and reduce project timelines. By optimizing material usage, construction techniques, and equipment utilization, these solutions can yield substantial savings in both time and resources, enhancing the overall competitiveness and profitability of construction projects.

Practical solutions derived from academic research have the potential to minimize environmental impact and promote sustainability. By incorporating green engineering principles, such as the use of recycled materials, energy-efficient construction methods, and sustainable land use practices, these solutions can reduce carbon emissions, conserve natural resources, and protect ecosystems. In doing so, they contribute to the long-term health and resilience of the built environment while mitigating the adverse effects of climate change.

To realize the full potential of academic research in driving practical solutions, it is essential to foster collaboration and knowledge exchange between researchers, practitioners, and industry stakeholders. This can be achieved through initiatives such as joint research projects, technology transfer programs, and professional development opportunities that facilitate the exchange of ideas, expertise, and best practices. By reducing the barriers between academia and industry, we can accelerate the pace of innovation, drive economic growth, and create a more sustainable future for generations to come.

In recent years, the field of geotechnical engineering has witnessed the emergence of promising new topics that hold significant potential for revolutionizing traditional practices and addressing contemporary challenges. Among these, ground improvement techniques and geothermal piles stand out as particularly promising areas of exploration.

Ground improvement techniques encompass a variety of methods aimed at enhancing the engineering properties of soils to meet the requirements of construction projects. These techniques offer alternative solutions to mitigate soil-related challenges such as low bearing capacity, excessive settlement, or poor drainage. Methods such as soil compaction, vibro-compaction, deep soil mixing, and grouting have been increasingly utilized to improve soil strength, stability, and drainage characteristics. Ground improvement techniques not only optimize the performance of foundation systems but also offer sustainable alternatives to traditional soil stabilization methods, reducing the need for excessive excavation and the use of non-renewable materials [5].

Geothermal piles represent another innovative approach gaining traction in geotechnical engineering. These specialized foundation elements integrate geothermal heat exchange systems into traditional pile foundations, allowing for the extraction or injection of thermal energy from or into the ground. By harnessing the stable thermal properties of the subsurface, geothermal piles offer opportunities for sustainable heating and cooling solutions in buildings and infrastructure projects. Beyond providing structural support, geothermal piles contribute to energy efficiency, reducing reliance on fossil fuels and lowering carbon emissions associated with heating, ventilation, and air conditioning systems [12, 13].

As these new topics continue to evolve, there is growing interest in their application across a wide range of construction projects, from high-rise buildings and transportation infrastructure to renewable energy installations and sustainable development initiatives. However, their successful implementation requires interdisciplinary collaboration, technical expertise, and rigorous performance evaluation to ensure reliability, efficiency, and environmental sustainability [14].

5. Artificial intelligence revolution

In the realm of geotechnical and foundation engineering, the fusion of scientific inquiry with practical experience is essential for tackling the evolving challenges faced by the construction industry while safeguarding the environment. This entails directing research and development efforts toward addressing practical challenges encountered in the field, such as soil stabilization, foundation design optimization, and construction material efficiency. Collaborative partnerships between academia, industry, and government agencies are instrumental in channeling resources toward research initiatives that yield tangible benefits for the construction sector and society at large. The emerging artificial intelligence, *AI* revolution, holds immense promise for revolutionizing the geotechnical industry. Advancements in machine learning, data analytics, and predictive modeling offer unprecedented opportunities to enhance accuracy, efficiency, and innovation across the entire project lifecycle. By leveraging AI algorithms to analyze vast datasets, simulate complex scenarios, and optimize design parameters, engineers can make more informed decisions, mitigate risks, and optimize resource utilization, ultimately leading to more sustainable and resilient infrastructure solutions [15].

Nonetheless, it is crucial to ensure that AI technologies are deployed ethically and responsibly, with due consideration given to issues of data privacy, algorithmic bias, and societal impact. Collaboration between AI experts, domain specialists, and stakeholders is key to developing AI-driven solutions that not only optimize performance but also uphold ethical principles and promote equitable outcomes.

Through this holistic approach that integrates scientific research, practical experience, and cutting-edge technologies, the geotechnical and foundation engineering community can drive innovation, sustainability, and resilience in the face of evolving challenges, paving the way toward a more sustainable and prosperous future for generations to come.

Author details

Mohamed Ayeldeen[1,2]

1 Arcadis Consulting, London, United Kingdom

2 Higher Institute of Engineering and Technology (THIET), Tanta, Egypt

*Address all correspondence to: ayeldeen.moh@gmail.com

References

[1] Bowles JE, Guo Y. Foundation Analysis and Design. Vol. 5. New York: McGraw-Hill; 1996. p. 127

[2] Poulos HG, Davis EH. Pile Foundation Analysis and Design. Vol. 397. New York: Wiley; 1980

[3] William Powrie. Soil Mechanics: Concepts and Applications. 2018

[4] Ayeldeen M, Negm A, El Sawwaf M, Gädda T. Laboratory study of using biopolymer to reduce wind erosion. International Journal of Geotechnical Engineering. 4 May 2018;**12**(3):228-240

[5] Ayeldeen M et al. Enhancing mechanical behaviors of collapsible soil using two biopolymers. Journal of Rock Mechanics and Geotechnical Engineering. 2017;**9**(2):329-339

[6] Ayeldeen M, Kitazume M. Using fiber and liquid polymer to improve the behaviour of cement-stabilized soft clay. Geotextiles and Geomembranes. 2017;**45**(6):592-602

[7] Azzam W, Ayeldeen M, El Siragy M. Improving the structural stability during earthquakes using in-filled trench with EPS geofoam—Numerical study. Arabian Journal of Geosciences. Jul 2018;**11**(14):1

[8] Basu D, Misra A, Puppala AJ. Sustainability and geotechnical engineering: Perspectives and review. Canadian Geotechnical Journal. 2014;**52**(1):96-113

[9] Jefferis SA. Moving Towards Sustainability in Geotechnical Engineering. In: GeoCongress 2008: Geosustainability and Geohazard Mitigation; 2008. pp. 844-851

[10] Lee M, Basu D. An integrated approach for resilience and sustainability in geotechnical engineering. Indian Geotechnical Journal. 2018;**48**(2): 207-234

[11] Morgenstern NR. Geotechnical engineering and frontier resource development. Geotechnique. 1981;**31**(3):305-365

[12] Cunha RP, Bourne-Webb PJ. A critical review on the current knowledge of geothermal energy piles to sustainably climatize buildings. Renewable and Sustainable Energy Reviews. 2022;**158**:112072

[13] Ghasemi-Fare O, Basu P. A practical heat transfer model for geothermal piles. Energy and Buildings. 2013;**66**:470-479

[14] De Moel M et al. Technological advances and applications of geothermal energy pile foundations and their feasibility in Australia. Renewable and Sustainable Energy Reviews. 2010;**14**(9):2683-2696

[15] Baghbani A et al. Application of artificial intelligence in geotechnical engineering: A state-of-the-art review. Earth-Science Reviews. 2022;**228**:103991

Chapter 2

Dynamic Stability of Beam on Elastic Foundation Including Higher Transition Foundation

Bhavanasi Subbaratnam

Abstract

A precise analytical solution has been developed using Energy Method to predict the dynamic instability bounds of simply supported beams on elastic foundation, with a focus on the impact of the higher transition foundation on dynamic stability boundaries. To determine the dynamic instability zones, a trigonometric function with a single term that meets the geometric boundary criteria to represent lateral deflection is used. For the analysis, Euler-Bernoulli beam theory is employed. Numerical results are presented in non-dimensional form in both digital and/or analogue forms, with varying foundation parameters below and above the transition foundation values of an elastic foundation. When compared to those produced using the finite element approach, the current findings exhibit a fair degree of consistency. The impact of the elastic foundation's first and higher transition foundation values on dynamic stability behavior is amply demonstrated in the current study. According to the studies, when the elastic foundation parameter value increases, the width of the dynamically unstable zones decreases, making the beam less susceptible to the dynamic stability phenomena under periodic loads. By using the precise non-dimensional parameters for the imposed periodic load and its radian frequency, the presence of the master dynamic instability curves is demonstrated in the current work.

Keywords: dynamic instability, uniform beams, energy method, periodic loads, elastic foundation, transition foundation

1. Introduction

An essential input for the structural design engineers is the prediction of the dynamic stability bounds of structural components exposed to periodic axial or in-plane loads. In his seminal work, Bolotin [1] covers in detail the dynamic instability of structural components/elements exposed to axial or in-plane periodic loads. In Ref. [2], a finite element technique is used to explore the dynamic instability of thin bars that are periodically exposed to intense axial loads at the free ends. In this paper, it is demonstrated intuitively that, provided the stability and vibration mode forms are same, the dynamic stability bounds derived with adequate non-dimensional parameters are the equal regardless of the boundary conditions. In [3], it is suggested that these non-dimensional parameters may be rigorously determined, and for the majority of structural members, it is shown that there are master dynamic instability curves that are valid for all structural members, regardless of boundary condition or complicating effects. If the requirement of the exactness of the mode shapes, but not the similarity as mentioned in [2], is broken while trying to get the master dynamic stability curves, the error involved in the analysis is dependent on the deviation of these mode shapes and may be evaluated by calculating the corresponding L_2 norms [4].

In several engineering areas, structural components like homogenous plates and beams on elastic foundations are frequently used. In Ref. [5], the impact of an elastic base on the dynamic stability of columns is examined. According to an instability discovery in Ref. [5], the zones of dynamic stability move away from the vertical axis and become narrower as the elastic foundation parameter rises, making the beam less susceptible to periodic loads. It has been demonstrated in Ref. [2] that the value of the foundation stiffness parameter affects the mode forms of columns under on elastic foundation. There is a value for transition foundation stiffness, and once this value is exceeded, the mode form of the buckling changes. For instance, when the foundation parameter is less than the initial/first transitional foundational parameter in the case of a simply supported beam, the stability and vibration mode forms are the equal (half of a sine wave along the length). When the foundation parameter is bigger than the initial transition value, these mode forms are different (full sinusoidal waves for the stability and half sinusoidal wave for the free vibration problems, respectively). The foundational parameters for this transition are discovered to be 4, 36, 144 ... for (first, second, third ...) for a simply supported beam/column [6]. Furthermore, it is demonstrated that the vibration mode forms of uniform beams supported by an elastic foundation do not correspond to such a transition foundation parameter. In Ref. [7], similar tests on rectangular plates with simple supports and biaxial periodic compressive stress on a homogenous elastic basis were described.

A beam that is axially loaded and rests on an elastic base with dampening is examined for dynamic stability in Ref. [8]. It has been demonstrated that raising the damping or stiffness of the foundation raises the critical dynamic load and moves the unstable zones to a higher/greater applied frequency. The dynamic instability of a tapered cantilever beam on an elastic basis was studied by Lee [9]. The dynamic instability of beams on elastic basis was explored by Subba Ratnam et al. [10]. By taking into account the reference buckling and frequency characteristics, Subba Ratnam et al. recently explored the dynamic instability of structural elements with secondary effects [11–15].

Wachirawit et al. [16] applied the Ritz and Newmark techniques for the Timoshenko beams to study the free vibrations and dynamic behavior of the functionally graded sandwich lying on an elastic base/foundation. Based on the Floquet theory, Fourier series, and matrix eigenvalue analysis, Ying et al. [17] studied multi-mode coupled periodically supported beams vibrating under generic harmonic excitations. Subbaratnam et al. [18] used the variational method to analyze the impact of dynamic instability boundaries of SS beams resting on elastic foundation under periodic loads caused by higher transition foundation. Using the Runge-Kutta method and the Floquet theorem, Chao Xu et al. [19] investigated dynamic instability zones of a simply supported beam under multi-harmonic parametric excitation Using the single matrix approach, Jian Deng et al. [20] investigated the dynamic stability and reactions of beams on elastic foundations under pulsing axial parametric load. The study and consequences of numerous factors, such as foundation basis models, damping, and static and dynamic loads, are taken into account using the Winkler, Pasternak, and Hetenyi models. Youqin Huang et al. [21] based on Reddy's beam theory, the dynamic instability of nanobeams was studied.

This study's major contribution is to examine how higher transition foundations affect the dynamic instability areas of SS beams lying on elastic foundations while being subjected to axial periodic loads. This study highlights the ease of employing a particular type of non-dimensional parameters employed, in order to show the results, and discusses the influence of altering mode forms for stability and vibration on the dynamic stability zones, with rising values of foundation parameter. This study demonstrates that the breadth of the zones of dynamic stability diminishes as the foundation increases, making a beam less responsive to the dynamic instability phenomena under periodic loads.

2. Formulation

Here, we briefly describe the mathematical formulation for SS beam on elastic foundation based on the variational principle and the assessment of the dynamic instability bounds.

When a regular beam of length L rests on elastic basis that is also uniformly distributed, it is subject to a end-axial periodic load P(t)., as shown in the **Figure 1**, the potential energy ($\prod$) is given, by

$$\prod = U + U_F - W - T \tag{1}$$

where T is kinetic energy, W is work produced by the external axial periodic load, U_F is the energy stored in the elastic foundation, and U is the strain energy. The formulas for U, U_F, T, and W are provided by

$$U = \frac{EI}{2}\int_0^L \left(\frac{d^2w}{dx^2}\right)^2 dx \tag{2}$$

$$U_F = \frac{k}{2}\int_0^L w^2 dx \tag{3}$$

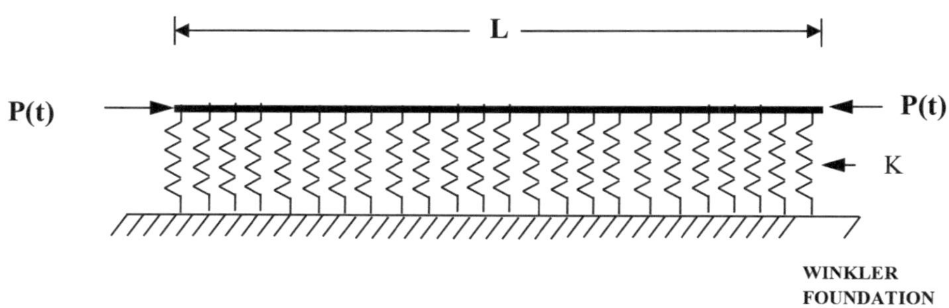

Figure 1.
Uniform SS beam lying on foundation subjected to periodic load.

$$T = \frac{m\omega^2}{2}\int_0^L w^2 dx \tag{4}$$

and

$$W = \frac{P(t)}{2}\int_0^L \left(\frac{dw}{dx}\right)^2 dx \tag{5}$$

where E stands for Modulus of Elasticity, I for moment of inertia, m is the mass per unit length, w for lateral displacement, P(t) for axial compressive periodic load, and ω is the natural radian frequency.

P(t) represents

$$P(t) = P_S + P_t \cos\theta t \tag{6}$$

where P_S is the periodic component of P(t), Pt is its periodic component, and θ is the compressive load's radian frequency. In terms of the buckling/critical load parameter Pcr, the numbers Ps and Pt are stated.

The differential equation controlling the dynamic stability problem of beams on elastic basis is expressed in terms of energy by substituting Eqs. (2)–(5) in Eq. (1).

$$\Pi = \frac{EI}{2}\int_0^L \left(\frac{d^2w}{dx^2}\right)^2 dx + \frac{k}{2}\int_0^L w^2 dx - \frac{P(t)}{2}\int_0^L \left(\frac{dw}{dx}\right)^2 dx - \frac{m\omega^2}{2}\int_0^L w^2 dx \tag{7}$$

As ω = θ/2, Eq. (7) becomes

$$\Pi = \frac{EI}{2}\int_0^L \left(\frac{d^2w}{dx^2}\right)^2 dx + \frac{k}{2}\int_0^L w^2 dx - \frac{P(t)}{2}\int_0^L \left(\frac{dw}{dx}\right)^2 dx - \frac{m}{2}\frac{\theta^2}{4}\int_0^L w^2 dx \tag{8}$$

When we replace the expression for P(t) from Eq. (6) in Eq. (8), we obtain

$$\Pi = \frac{EI}{2}\int_0^L \left(\frac{d^2w}{dx^2}\right)^2 dx + \frac{k}{2}\int_0^L w^2 dx - (P_S + P_t \cos\theta t)\frac{1}{2}\int_0^L \left(\frac{dw}{dx}\right)^2 dx - \frac{m}{2}\frac{\theta^2}{4}\int_0^L w^2 dx \tag{9}$$

The boundary between Eq. (9)'s stable and unstable solutions are periodic solutions with periods of 2π/θ and 4π/θ according to the theory of linear equations with periodic coefficients [1]. The unstable solutions that are constrained by the solutions with period's 4π/θ are those that have the most practical significance. The need for these boundary solutions, as a first approximation, is

$$\Pi = \frac{EI}{2}\int_0^L \left(\frac{d^2w}{dx^2}\right)^2 dx + \frac{k}{2}\int_0^L w^2 dx - \left(\alpha \pm \frac{\beta}{2}\right)\frac{P_{cr}}{2}\int_0^L \left(\frac{dw}{dx}\right)^2 dx - \frac{m}{2}\frac{\theta^2}{4}\int_0^L w^2 dx \tag{10}$$

The boundary between the stable and unstable solutions of Eq. (8) are periodic solutions with periods of 2π/θ and 4π/θ according to the theory of linear equations with periodic coefficients [1].

where α is the fraction of buckling load $(=\frac{P_s}{P_{cr}})$ and β is the fraction of periodic buckling load

$$\left(=\frac{P_t}{P_{cr}}.\right) \tag{11}$$

It should be noted that Eq. (10), which yields the two borders of the zones of instability, combines two requirements in the plus or minus sign. The beam's static buckling load factor is used as the standard in this research.

2.1 Buckling (critical) load parameter and frequency parameter of SS beam on elastic foundation

The fundamental equation for dynamic stability provides the corresponding solutions for the Euler buckling loads and frequency parameters as functions of the wave number n, taking into account the influence of the elastic foundation. This is done by non-dimensionalizing all length quantities by the length of the beam L and neglecting the kinetic energy term (T) and work done by the periodic load (W), respectively.

Using the single term standard exact trigonometric admissible function, for a SS beam, for lateral deflection, as

$$w = a \sin \frac{n\pi x}{L} \tag{12}$$

where a is the indeterminate coefficient, n is the mode shape number, (denoted by n_b for the buckling issue and n_f for the free vibration problem), w is the lateral deflection, and x is the axial coordinate, respectively.

In terms of the mode numbers n_b and n_f, this admissible function provides the buckling loads and the frequency parameters as

$$\lambda_b = \frac{P_{cr}L^2}{EI} = \pi^2\left(n^2 + \frac{\gamma}{n^2}\right) \tag{13}$$

and

$$\lambda_f = \frac{m\omega^2 L^4}{EI} = \pi^4\left(n^2 + \frac{\gamma}{n^2}\right) \tag{14}$$

where γ is the foundation parameter $(\gamma = \frac{kL^4}{\pi^4 EI})$, λ_b is buckling/critical load parameter $(=\frac{P_{cr}L^2}{EI})$ and λ_f is frequency parameter $(=\frac{m\omega^2 L^4}{EI})$.where P_{cr} is critical load and ω is natural frequency.

2.2 Foundation parameter transition values

The transitional foundation parameter value γ_{Ti} (i varies from 1, 2, 3 … . and with reference to buckling mode number n_b), where buckling load parameter λ_b changes from mode n_b to mode $(n + 1)_b$. The transition value γ_{Ti} is evaluated from Eq. (13), using the condition that the buckling load parameters are the same for the two consecutive modes of buckling n_b and $(n + 1)_b$, as

$$\lambda_{bn} = \lambda_{b(n+1)} \tag{15}$$

The values of the γ_{Ti} are

$$\gamma_{Ti} = {n_b}^2 (n+1)_{\mathrm{b}}^{\ 2}, (i, n_b = 1, 2, 3, 4, 5 \dots.) \tag{16}$$

Two noteworthy findings from the assessment of the γ_{Ti} are:

a. the contribution of γ_{Ti} for buckling problem,

b. a similar equation as Eq. (16) for the free vibration issue, where the mode forms vary, the phenomena of transition does not occur.

For buckling issue, the values of γ_{Ti} are 4, 36, 144 ... for (i = 1, 2, 3,4 ...).
The impact of the initial and second transition foundation parameters γ_{T1} = 4 and γ_{T2} = 36, respectively and its impact on dynamic instability behavior of SS beam are both thoroughly explored in the current work.

2.3 Variation of λ_b and λ_f for different values of n_b and n_f

Eqs. (13) and (14), provide expressions for λ_b and λ_f the various values of n_b, n_f and γ. In light of these values of γ, n_b and n_f, succinct evaluations of the formulations for λ_b and λ_f are given here;

2.3.1 For $\gamma < \gamma_{T1}$ ($n_b = n_f = 1$)

The λ_b and λ_f terms are

$$\lambda_b = \frac{P_{cr} L^2}{EI} = \pi^2 [1 + \gamma] \tag{17}$$

and

$$\lambda_f = \frac{m\omega^2 L^4}{EI} = \pi^4 (1 + \gamma) \tag{18}$$

2.3.2 For $\gamma > \gamma_{T1}$ and $\gamma < \gamma_{T2}$ ($n_b = 2$, $n_f = 1$)

The λ_b and λ_f terms are

$$\lambda_b = \frac{P_{cr} L^2}{EI} = \pi^2 \left[4 + \frac{\gamma}{4}\right] \tag{19}$$

and

$$\lambda_f = \frac{m\omega^2 L^4}{EI} = \pi^4 (1 + \gamma) \tag{20}$$

2.3.3 For $\gamma > \gamma_{T1}$ and $\gamma < \gamma_{T2}$ ($n_b = 2$, $n_f = 2$)

The λ_b and λ_f terms are

$$\lambda_b = \frac{P_{cr}L^2}{EI} = \pi^2\left[4 + \frac{\gamma}{4}\right] \tag{21}$$

and

$$\lambda_f = \frac{m\omega^2 L^4}{EI} = \pi^4(16 + \gamma) \tag{22}$$

2.3.4 For $\gamma > \gamma_{T2}$ and $\gamma < \gamma_{T3}$ ($n_b = 3$, $n_f = 1$)

The λ_b and λ_f terms are

$$\lambda_b = \frac{P_{cr}L^2}{EI} = \pi^2\left[9 + \frac{\gamma}{9}\right] \tag{23}$$

and

$$\lambda_f = \frac{m\omega^2 L^4}{EI} = \pi^4(1 + \gamma) \tag{24}$$

2.3.5 For $\gamma > \gamma_{T2}$ and $\gamma < \gamma_{T3}$ ($n_b = 3$, $n_f = 3$)

The λ_b and λ_f terms are

$$\lambda_b = \frac{P_{cr}L^2}{EI} = \pi^2\left[9 + \frac{\gamma}{9}\right] \tag{25}$$

and

$$\lambda_f = \frac{m\omega^2 L^4}{IE} = \pi^4(81 + \gamma) \tag{26}$$

For the sake of completeness, the expressions for λ_b and λ_f for the three cases that are being considered here are given, but any combination of n_b, n_f and γ, the buckling and frequency parameters can be immediately obtained for any given value of these four parameters by taking into account transition value of foundation parameter for γ.

3. Formulas for dynamic stability

The following formulae are short and to the point for forecasting the dynamic instability areas of the simply supported beam on the foundation:

The dynamic stability formula is determined by replacing Eqs. (12), (17), and (18) in Eq. (10) and utilizing the first mode of buckling and free vibration (nb = nf = 1) as the reference values, where $P_{cr} = \frac{\pi^2 EI}{L^2}$ and $m = \frac{\pi^4 EI}{\omega^2 L^4}$ without taking into account the impact of foundation.

$$1 + \gamma - \left[\alpha \pm \frac{\beta}{2}\right] - \frac{\theta^2}{4\omega^2} = 0 \tag{27}$$

$$\alpha = \frac{P_s}{P_{cr}} \text{ and } \beta = \frac{P_t}{P_{cr}}. \tag{28}$$

or

$$\frac{\theta}{\omega} = \Omega = 2\sqrt{(1-\alpha)(1\pm\mu) + \gamma} \tag{29}$$

where μ is the defined non-dimensional parameter, $\mu = \frac{\beta}{2(1-\alpha)}$ used in some earlier studies on the topic of dynamic stability [2].

The dynamic stability formula is formulated as follows for n_b = 2 and n_f = 1 when Eqs. (12), (19), and (20) are substituted in Eq. (10)

$$1 - \frac{\left(\alpha \pm \frac{\beta}{2}\right)P_{cr}.4\frac{\pi^2}{L^2}}{\frac{\pi^4 EI}{L^4}\left[16 + \frac{kL^4}{\pi^4 EI}\right]} - \frac{\frac{m\theta^2}{4}}{\frac{\pi^4 EI}{L^4}\left[1 + \frac{kL^4}{\pi^4 EI}\right]} = 0 \tag{30}$$

Substituting the corresponding reference values of $P_{cr} = \frac{4\pi^2 EI}{L^2}$ and $m = \frac{\pi^4 EI}{\omega^2 L^4}$ without considering the effect of the elastic foundation in Eq. (30) becomes

$$\Omega = \frac{\theta}{\omega} = 2\sqrt{\left[\left((1-\alpha)(1\pm\mu) + \frac{\gamma}{16}\right)\left(\frac{(1+\gamma)}{\left(1+\frac{\gamma}{16}\right)}\right)\right]} \tag{31}$$

In the same way, the dynamic stability formula is as follows when Eqs. (12), (23), and (24), respectively, are substituted for n_b = 3 and n_f = 1 in Eq. [10], as

$$1 - \frac{\left(\alpha \pm \frac{\beta}{2}\right)P_{cr}.9\frac{\pi^2}{L^2}}{\frac{\pi^4 EI}{L^4}\left[81 + \frac{kL^4}{\pi^4 EI}\right]} - \frac{\frac{m\theta^2}{4}}{\frac{\pi^4 EI}{L^4}\left[1 + \frac{kL^4}{\pi^4 EI}\right]} = 0 \tag{32}$$

Eq. (32) is expressed as follows when the reference values $P_{cr} = \frac{9\pi^2 EI}{L^2}$ and $m = \frac{\pi^4 EI}{\omega^2 L^4}$ are substituted and the foundation is ignored.

$$\Omega = \frac{\theta}{\omega} = 2\sqrt{\left[\left[(1-\alpha)(1\pm\mu) + \frac{\gamma}{81}\right]\left[\frac{(1+\gamma)}{(1+(\gamma/81))}\right]\right]} \tag{33}$$

These three dynamic instability formulae are constructed using the modes n_b and n_f for initial and second transition foundation parameter values γ smaller or greater than γ_{T1} and γ_{T2}.

4. Master dynamic stability formula

Interestingly, the master dynamic stability formula can be obtained by substituting Eq. (12) and the reference buckling load Pcr and frequency ω into Eq. (10), along with half sinusoidal wave for both buckling and vibration problems (nb = nf = 1), a full sinusoidal wave for both buckling and vibration problems (nb = nf = 2), and three half sinusoidal waves for both the stability and vibration problems (nb = nf = 3), as

$$\frac{\theta}{\omega} = \Omega = 2\sqrt{(1-\alpha)(1 \pm \mu)} \tag{34}$$

Use of these specific reference values of λ_b and λ_f the master dynamic stability curves turn out to be simple to use, exactly the same and are independent of the foundation parameter for the same values of n_b and n_f (either 1, 2 or 3). It should be noticed that, unlike preceding formulations where the foundation parameters appeared clearly, the master dynamic stability formula shown in Eq. (34) does not explicitly include the foundation parameter (γ).

5. Results and discussion

A homogeneous simply supported beam that is sitting on elastic basis and being subjected to axial concentrated periodic load is shown in **Figure 1**. The equation created in the current work to examine the dynamic instability limits of a SS beam sitting on an elastic basis and subjected to end axial periodic concentrated load is highly universal. The fundamental frequency and the static buckling load parameters, which are characteristic values of the beam, do not clearly exist in the non-dimensional form. However, when non-dimensional parameters μ and Ω. are defined, their characteristic values are referenced implicitly.

The values of the stability boundaries $\Omega_1$and Ω_2, between which it is dynamically unstable with variable for α = 0.0, 0.5 & 0.8 with mode form of a half sinusoidal wave for both stability and vibration problems for γ = 1, are shown in **Table 1**. **Table 2** gives the values of the stability boundaries $\Omega_1$and Ω_2 with varying β, instead of μ, for α = 0.0, 0.5 & 0.8 for mode form of one half sinusoidal wave for both stability & vibration problems for γ = 2. The dynamic instability results of a slender beam on foundation [5] are also included in this Table. Since the non-dimensional parameter given in Ref. [5] is β instead of μ proposed here, and as it is difficult to convert β to μ due to lack of information, the values of μ is converted in to β for the

μ	$\alpha = 0.0$		$\alpha = 0.5$		$\alpha = 0.8$	
	Ω_1	Ω_2	Ω_1	Ω_2	Ω_1	Ω_2
0.0	2.8284	2.8284	2.4494	2.4494	2.1908	2.1908
0.1	2.7568	2.8982	2.4083	2.4899	2.1725	2.2090
0.2	2.6832	2.9664	2.3664	2.5298	2.1540	2.2271
0.3	2.6076	3.0331	2.3237	2.5690	2.1354	2.2449
0.4	2.5298	3.0983	2.2803	2.6076	2.1166	2.2627
0.5	2.4494	3.1622	2.2360	2.6457	2.0976	2.2803

λ_b and λ_f values for $\gamma = 0$.

Table 1.
Variation of Ω_1 and Ω_2 for half sinusoidal waves in both stability and vibration for $\gamma = 1.0$ ($\gamma < \gamma_{T1}$)*.

μ	$\alpha = 0.0$		$\alpha = 0.5$		$\alpha = 0.8$	
	Ω_1	Ω_2	Ω_1	Ω_2	Ω_1	Ω_2
0.0	3.4641	3.4641	3.1622 (3.1439)$^{\$}$	3.1622 (3.1439)	2.9664	2.9664
0.2	3.4058	3.5213	3.0983 (3.0384)	3.2249 (3.1818)	2.8982	3.0331
0.4	3.3466	3.5777	3.0331 (3.0000)	3.2863 (3.2575)	2.8284	3.0983
0.6	3.2863	3.6331	2.9664 (2.9090)	3.3466 (3.3333)	2.7568	3.1622
0.8	3.2249	3.6878	2.8982 (2.8787)	3.4058 (3.4090)	2.6832	3.2249
1.0	3.1622	3.7416	2.8284 (2.8181)	3.4615 (3.4545)	2.6076	3.2863

$^{\$}$Values in the parenthesis are read from Ref. [5].$^{*}\lambda_b$ and λ_f values for $\gamma = 0$.

Table 2.
Variation of $\Omega_1$and Ω_2 for half sinusoidal waves in both stability and vibration for $\gamma = 2.0$ $(\gamma < \gamma_{T1})^{*}$.

μ	$\alpha = 0.0$		$\alpha = 0.5$		$\alpha = 0.8$	
	Ω_1	Ω_2	Ω_1	Ω_2	Ω_1	Ω_2
0.0	4.0000	4.0000	3.7416	3.7416	3.5777	3.5777
0.1	3.9496	4.0496	3.7148	3.7682	3.5665	3.5888
0.2	3.8987	4.0987	3.6878	3.7947	3.5552	3.6001
0.3	3.8471	4.1472	3.6606	3.8209	3.5441	3.6111
0.4	3.7947	4.1952	3.6331	3.8471	3.5327	3.6221
0.5	3.7416	4.2426	3.6055	3.8729	3.5213	3.6331

$^{*}\lambda_b$ and λ_f values for $\gamma = 0$.

Table 3.
Variation of $\Omega_1$and Ω_2 for half sinusoidal waves in both stability and vibration for $\gamma = 3.0$ $(\gamma < \gamma_{T1})^{*}$.

results presented in the Table. The results are in good accord with results obtained from present formula and those given in Ref. [5] obtained using finite approach for α = 0.5. **Table 3** gives the values of the stability boundaries $\Omega_1$and Ω_2, between which it exhibits dynamic instability with variable μ for α = 0.0, 0.5 and 0.8, and a half sinusoidal wave mode form for both stability and vibration issues γ = 3. The instability boundaries $\Omega_1$and Ω_2 given in **Tables 1–3** are for α = 0.0, 0.5 and 0.8 with γ = 1, 2 and 3 respectively. For a better understanding of the zones of dynamic stability, **Figures 2–4** illustrate the boundaries of the instability that are described in **Tables 1–3**. Additionally, it can be shown that the zones of dynamic instability shifts away from the vertical axis and reduce as the elastic foundation parameter increases, making the dynamic stability phenomena less sensitive to periodic loads.

Tables 4–7 gives the values of the stability boundaries $\Omega_1$and Ω_2, within which it is dynamically unstable with change μ for α = 0.6, 0.7 & 0.8 with the mode form of full sinusoidal wave for the stability problem and half sinusoidal wave for the vibration problem for γ > 4 $(\gamma > \gamma_{T1})$. The instability boundaries $\Omega_1$and Ω_2 given in **Tables 4–7** are for α = 0.6, 0.7 and 0.8 with γ = 5 to γ = 35. For an improved understanding of the zones of dynamic stability, **Figures 5–8** illustrate the instability boundaries $\Omega_1$and Ω_2

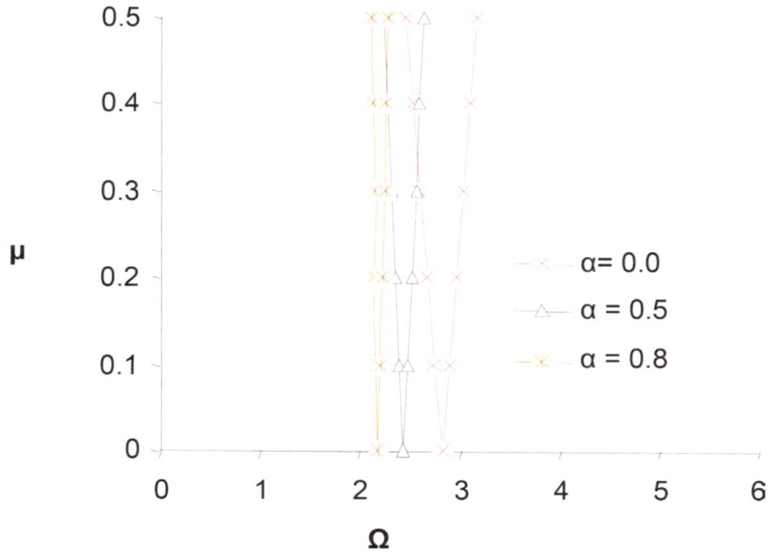

Figure 2.
Dynamic stability bounds for half sinusoidal wave in both stability and vibration for $\gamma = 1 (\gamma < 4)$.

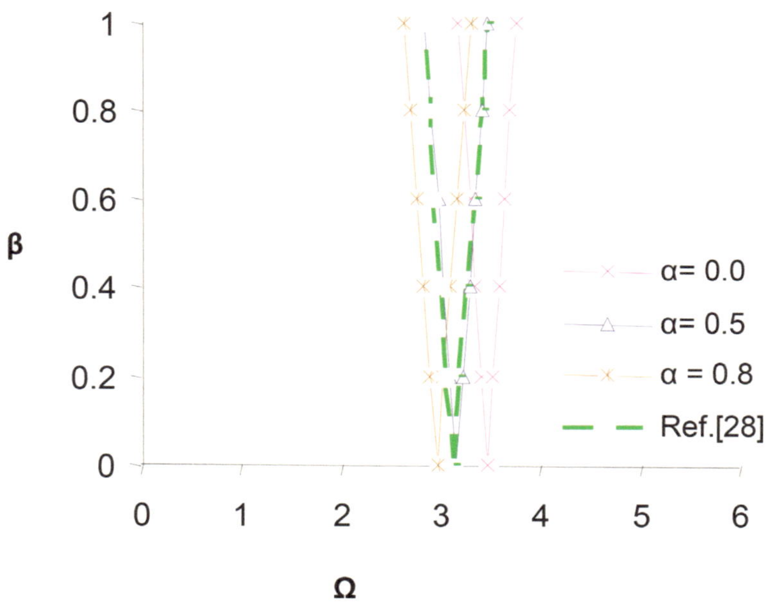

Figure 3.
Dynamic stability bounds for half sinusoidal wave in both stability and vibration for $\gamma = 2$ $(\gamma < 4)$.

provided in **Tables 4–7**. Additionally, it can be shown that when the foundation parameter increases, the zones of dynamic instability move away from the vertical axis. The areas of dynamic instability for γ = 5 to γ = 35.0 decrease from above the initial transition foundation to below second transition foundation parameter value, which is an important finding in this study.

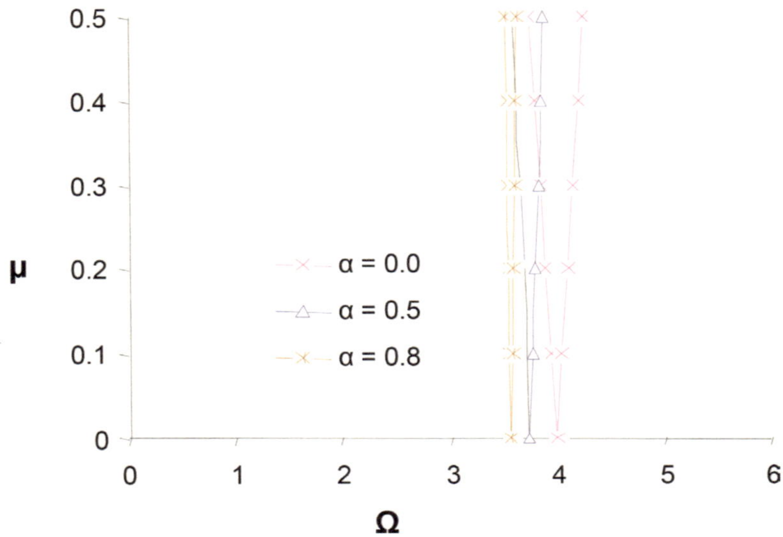

Figure 4.
Dynamic stability bounds for half sinusoidal wave in both stability and vibration for on for $\gamma = 3$ ($\gamma < 4$).

μ	$\alpha = 0.6$		$\alpha = 0.7$		$\alpha = 0.8$	
	Ω_1	Ω_2	Ω_1	Ω_2	Ω_1	Ω_2
0.0	3.6095	3.6095	3.3466	3.3466	3.0612	3.0612
0.1	3.5067	3.7094	3.2636	3.4276	3.0009	3.1204
0.2	3.4008	3.8067	3.1784	3.5067	2.9393	3.1784
0.3	3.2915	3.9016	3.0910	3.5840	2.8765	3.2355
0.4	3.1784	3.9942	3.0009	3.6598	2.8122	3.2915
0.5	3.0612	4.0848	2.9081	3.7340	2.7464	3.3466

*λ_b and λ_f values for $\gamma = 0$.

Table 4.
Variation of $\Omega_1$and Ω_2 for full sinusoidal wave for stability and half sinusoidal wave for vibration for $\gamma = 5.0$ ($\gamma > \gamma_{T1}$).*

μ	$\alpha = 0.6$		$\alpha = 0.7$		$\alpha = 0.8$	
	Ω_1	Ω_2	Ω_1	Ω_2	Ω_1	Ω_2
0.0	6.6468	6.6468	6.3935	6.3935	6.1297	6.1297
0.1	6.5467	6.7455	6.3155	6.4705	6.0756	6.1834
0.2	6.4450	6.8427	6.2366	6.5467	6.0210	6.2366
0.3	6.3416	6.9386	6.1566	6.6219	5.9659	6.2893
0.4	6.2366	7.0331	6.0756	6.6963	5.9102	6.3416
0.5	6.1297	7.1265	5.9935	6.7699	5.8541	6.3935

*λ_b and λ_f values for $\gamma = 0$.

Table 5.
Variation of $\Omega_1$and Ω_2 for full sinusoidal wave for stability and half sinusoidal wave for vibration for $\gamma = 15.0$ ($\gamma > \gamma_{T1}$).*

μ	$\alpha = 0.6$		$\alpha = 0.7$		$\alpha = 0.8$	
	Ω_1	Ω_2	Ω_1	Ω_2	Ω_1	Ω_2
0.0	8.9246	8.9246	8.6942	8.6942	8.4576	8.4576
0.1	8.8331	9.0151	8.6239	8.7640	8.4095	8.5054
0.2	8.7408	9.1047	8.5530	8.8331	8.3611	8.5530
0.3	8.6474	9.1934	8.4816	8.9018	8.3124	8.6003
0.4	8.5530	9.2812	8.4095	8.9699	8.2634	8.6474
0.5	8.4576	9.3683	8.3368	9.0375	8.2142	8.6942

*λ_b and λ_f values for $\gamma = 0$.

Table 6.
Variation of Ω_1 and Ω_2 for full sinusoidal wave for stability and half sinusoidal wave for vibration for $\gamma = 25.0$ $(\gamma > \gamma_{T1})^*$.

μ	$\alpha = 0.6$		$\alpha = 0.7$		$\alpha = 0.8$	
	Ω_1	Ω_2	Ω_1	Ω_2	Ω_1	Ω_2
0.0	10.8117	10.8117	10.6007	10.6007	10.3855	10.3855
0.1	10.7278	10.8950	10.5366	10.6645	10.3419	10.4289
0.2	10.6433	10.9776	10.4721	10.7278	10.2981	10.4721
0.3	10.5580	11.0596	10.4072	10.7908	10.2541	10.5151
0.4	10.4721	11.1410	10.3419	10.8534	10.2100	10.5580
0.5	10.3855	11.2218	10.2761	10.9157	10.1656	10.6007

*λ_b and λ_f values for $\gamma = 0$.

Table 7.
Variation of Ω_1 and Ω_2 for full sinusoidal wave for stability and half sinusoidal wave for vibration for $\gamma = 35.0$ $(\gamma > \gamma_{T1})^*$.

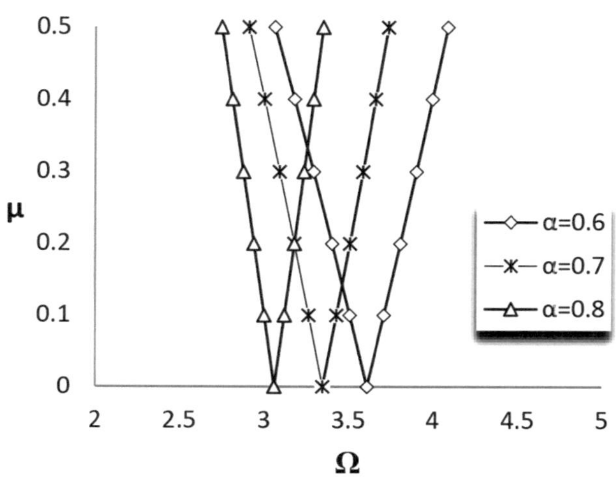

Figure 5.
Dynamic stability bounds for full sinusoidal wave for stability and half sinusoidal wave for vibration for $\gamma = 5.0$ $(\gamma > \gamma_{T1})$.

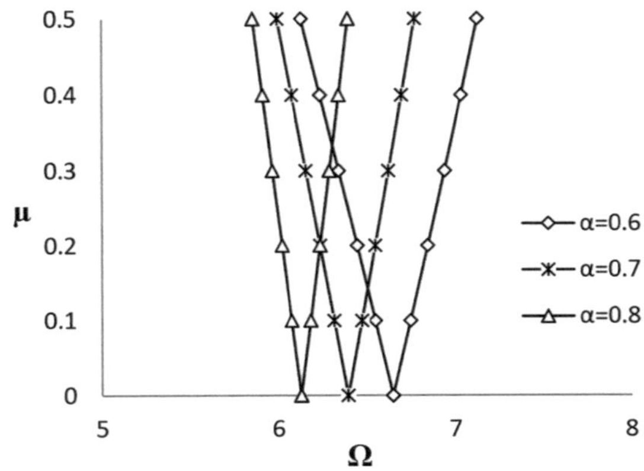

Figure 6.
Dynamic stability curves for full sinusoidal wave for stability and half sinusoidal wave for vibration for $\gamma = 15.0$ $(\gamma > \gamma_{T1})$.

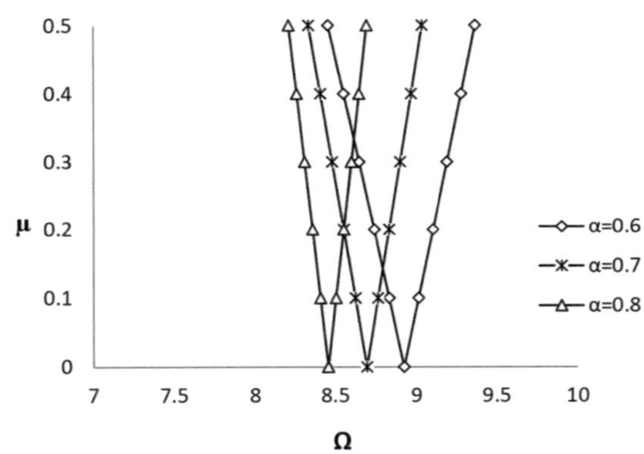

Figure 7.
Dynamic stability curves for full sinusoidal wave for stability and half sinusoidal wave for vibration for $\gamma = 25.0$ $(\gamma > \gamma_{T1})$.

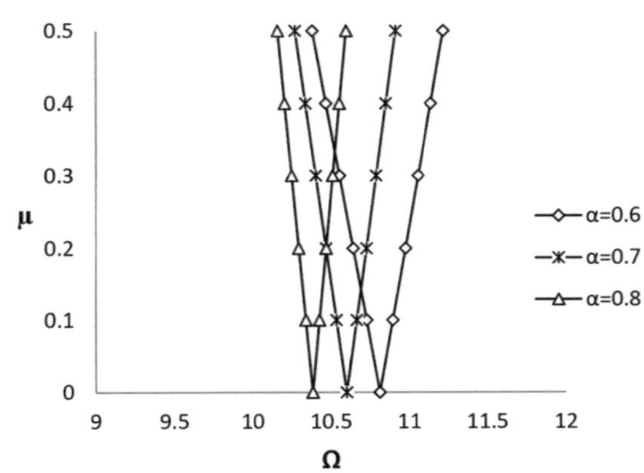

Figure 8.
Dynamic stability curves for full sinusoidal wave for stability and half sinusoidal wave for vibration for $\gamma = 35.0$ $(\gamma > \gamma_{T1})$.

μ	$\alpha = 0.6$		$\alpha = 0.7$		$\alpha = 0.8$	
	Ω_1	Ω_2	Ω_1	Ω_2	Ω_1	Ω_2
0.0	10.6314	10.6314	10.0598	10.0598	9.4536	9.4536
0.1	10.4065	10.8517	9.8818	10.2346	9.3276	9.5779
0.2	10.1767	11.0675	9.7006	10.4065	9.2000	9.7006
0.3	9.9415	11.2793	9.5160	10.5757	9.0705	9.8218
0.4	9.7006	11.4871	9.3276	10.7421	8.9391	9.9415
0.5	9.4536	11.6912	9.1354	10.9060	8.8058	10.0598

*λ_b and λ_f values for $\gamma = 0$.

Table 8.
Variation of Ω_1 and Ω_2 for three half sinusoidal waves for stability and half sinusoidal wave for vibration for $\gamma = 45.0$ $(\gamma > \gamma_{T2})^*$.

μ	$\alpha = 0.6$		$\alpha = 0.7$		$\alpha = 0.8$	
	Ω_1	Ω_2	Ω_1	Ω_2	Ω_1	Ω_2
0.0	11.9980	11.9980	11.4285	11.4285	10.8291	10.8291
0.1	11.7735	12.2184	11.2520	11.6023	10.7052	10.9516
0.2	11.5446	12.4348	11.0728	11.7735	10.5798	11.0728
0.3	11.3112	12.6476	10.8905	11.9423	10.4530	11.1926
0.4	11.0728	12.8568	10.7052	12.1087	10.3246	11.3112
0.5	10.8291	13.0627	10.5166	12.2728	10.1945757	11.4285

*λ_b and λ_f values for $\gamma = 0$.

Table 9.
Variation of Ω_1 and Ω_2 for three half sinusoidal waves for stability and half sinusoidal wave for vibration for $\gamma = 55.0$ $(\gamma > \gamma_{T2})^*$.

μ	$\alpha = 0.6$		$\alpha = 0.7$		$\alpha = 0.8$	
	Ω_1	Ω_2	Ω_1	Ω_2	Ω_1	Ω_2
0.0	12.6450	12.6450	12.0781	12.0781	11.4831	11.4831
0.1	12.4213	12.8648	11.9027	12.2509	11.3604	11.6046
0.2	12.1936	13.0809	11.7247	12.4213	11.2364	11.7247
0.3	11.9614	13.2935	11.5440	12.5895	11.1109	11.8437
0.4	11.7247	13.5027	11.3604	12.7554	10.9840	11.9614
0.5	11.4831	13.7088	11.1738	12.9192	10.8557	12.0781

*λ_b and λ_f values for $\gamma = 0$.

Table 10.
Variation of Ω_1 and Ω_2 for three half sinusoidal waves for stability and half sinusoidal wave for vibration for $\gamma = 60.0$ $(\gamma > \gamma_{T2})^*$.

Tables 8–10 provide the values of the stability boundaries $\Omega_1$and $\Omega_2$that, for three half-sinusoidal waves for stability problems and half-sinusoidal wave for vibration problems, respectively, are dynamically unstable above the value of the second transition foundation parameter ($\gamma > 36$) for $\gamma = 45.0$ to $\gamma = 60.0$. It is noted that the regions of dynamic instability are pushed away from the vertical axis and the instability zones are initially bigger above the second transition foundation value. Another noteworthy finding is that the dynamic instability areas for values above the second transition foundation and below the third transition foundation value decrease with increasing foundation value. For a better understanding of the limits/bounds of dynamic instability, **Figures 9–11** demonstrate the instability boundaries $\Omega_1$and Ω_2 provided in **Tables 8** and **10**. **Figure 12** displays the dynamic instability curves for full

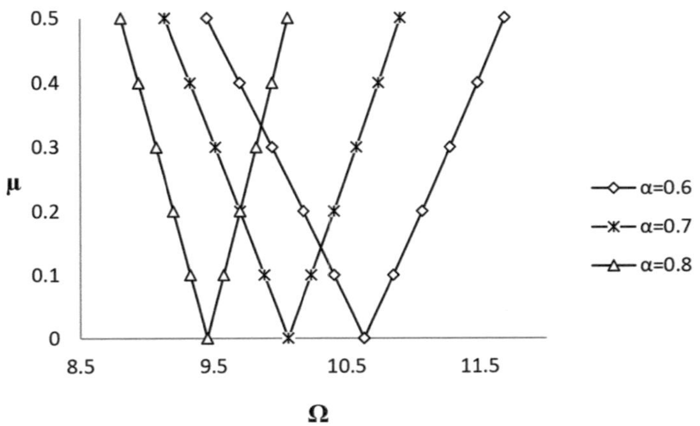

Figure 9.
Dynamic stability curves for three half sinusoidal waves for stability and half sinusoidal wave for vibration for $\gamma = 45.0$ $(\gamma > \gamma_{T2})$.

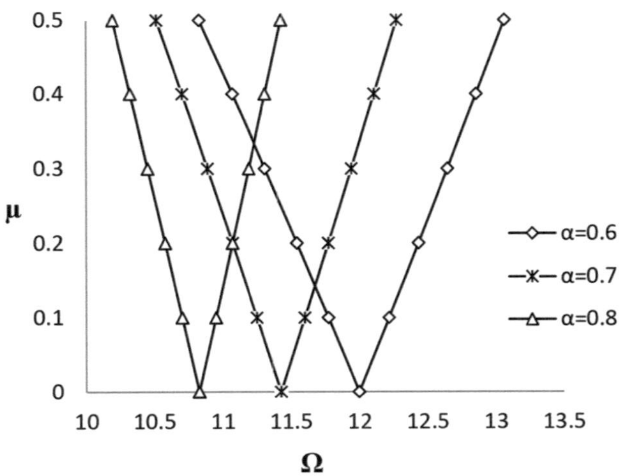

Figure 10.
Dynamic stability curves for three half sinusoidal waves for stability and half sinusoidal wave for vibration for $\gamma = 55.0$ $(\gamma > \gamma_{T2})$.

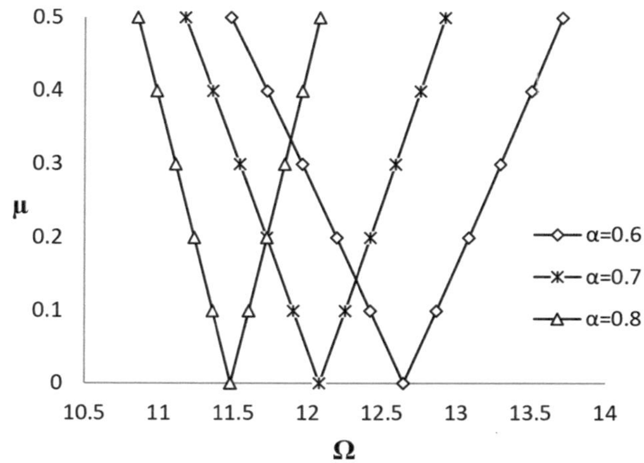

Figure 11.
Dynamic stability curves for three half sinusoidal waves for stability and half sinusoidal wave for vibration for $\gamma = 60.0$ ($\gamma > \gamma_{T2}$).

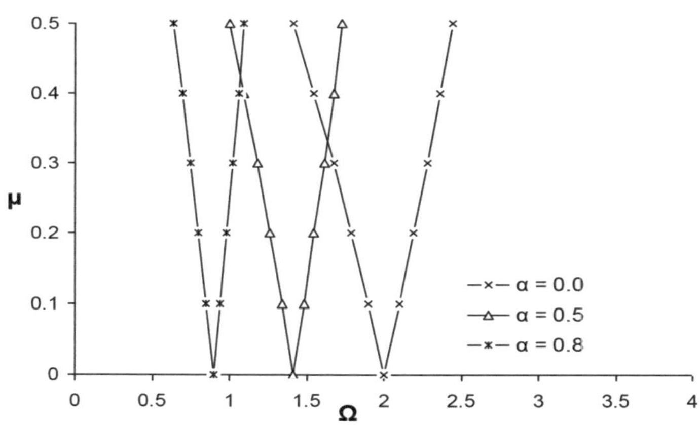

Figure 12.
Master dynamic stability curves.

sinusoidal waves for stability and vibration problems, with reference buckling load and reference frequency parameters taking into account elastic foundation for the foundation parameter below the second transition value ($\gamma < 36$), and three half-sinusoidal waves for stability and vibration problems, with reference buckling load and reference frequency parameters taking into account foundation for the foundation parameter above the second transition value ($\gamma > 36$) obtained from Eq. (34).

6. Conclusions

To accurately forecast the dynamic stability behavior of SS beam resting on elastic basis under periodic axial load, closed form solutions are derived. The geometric boundary criteria are satisfied by a one-term trigonometric admissible function. Suitable non-dimensional parameters come out of the present formulation following the Variational method that is the basis for the standard Rayleigh-Ritz technique. The existence of transition, where the mode form changes for the stability/buckling problem, is discussed. The numerical findings from the current formulation and those from the finite element approach are in good agreement. It is also proven how higher transition foundations affect the dynamic stability zones and borders. This study presents the findings obtained with different first and second foundation parameters, both below and above this transition threshold. The breadth/width of the zones of dynamic stability diminishes as elastic basis does, making the beam less susceptible to the dynamic stability phenomena under periodic loads. When the reference values of the buckling load and radian frequency are evaluated taking into consideration the effect of the foundation, which is not recognized by the earlier researchers, the existence of the master dynamic stability curves for the same mode numbers for the buckling and free vibration problems is established. The versatility of our proposed method is highlighted by its potential extension to the analysis of dynamic instability boundaries for different boundary conditions, such as Pasternak foundation, Timoshenko beams, and nanobeams, all subjected to axial periodic loads.

Acknowledgements

The author expresses gratitude to the management of Malla Reddy Engineering College and Management Sciences for their continuous support throughout this work.

Author details

Bhavanasi Subbaratnam
Department of Mechanical Engineering, Malla Reddy Engineering College and Management Sciences, Kistapur, Medchal, Hyderabad, India

*Address all correspondence to: b_subbaratnam@yahoo.com

References

[1] Bolotin VV. Dynamic Stability of Elastic Systems. San Francisco: Holden-Day; 1964

[2] Brown JE, Hutt JM, Salama AE. Finite element solution to dynamic stability of bars. AIAA Journal. 1968;**6**: 1423-1425

[3] Rao GV, Ratnam BS, Janardhana GR. Master dynamic stability formula for structural members subjected to periodic loads. AIAA Journal. 2008;**46**:537-540

[4] Rao GV, Ratnam BS, Gunda JB, Janardhana GR. Master formula for evaluating vibration frequencies of structural members under compressive loads. The IES Journal Part A: Civil & Structural Engineering. 2011;**4**(2):79-88

[5] Abbas BAH, Thomas J. Dynamic stability of Timoshenko beams resting on an elastic foundation. Journal of Sound and Vibration. 1978;**60**:33-44

[6] Timoshenko SP, Gere JM. Theory of Elastic Stability. New York: Mc Graw-Hill; 1936

[7] Kanaka Raju K, Venkatesara Rao G. Vibration of initially stressed beams and plates around transition values of elastic foundation stiffness. Journal of Sound and Vibration. 1993;**161**: 378-384

[8] Engel RS. Dynamic stability of axially loaded beam on an elastic foundation with damping. Journal of Sound and Vibration. 1991;**146**(3):463-477

[9] Lee HP. Dynamic stability of a tapered cantilever beam on an elastic foundation subjected to a follower force. International Journal of Solids Structures. 1996;**33**(10):1409-1424

[10] Ratnam BS, Rao GV, Janardhana GR. Dynamic stability of beams on elastic foundation. Journal of Structural Engineering. 2011;**38**(2):150-160

[11] Subba Ratnam B, Venkateswara Rao G, Ranga Janardhana G. Development of three simple master dynamic stability formulas for structural members subjected to periodic load. Journal of The Institution of Engineers (India). 2011;**92**:9-14

[12] Subba Ratnam B, Venkateswara Rao G, Ranga Janardhana G. Dynamic stability of beams subjected to end periodic and static tensile axial load. Journal of Structural Engineering. 2012; **39**(2):263-268

[13] Subba Ratnam B, Venkateswara Rao G. Prediction of dynamic stability behavior of structural members subjected to periodic load through the concept of dynamic stability point. Journal of Structural Engineering. 2014; **41**(5):559-564

[14] Subba Ratnam B, Venkateswara Rao G, Ranga Janardhana G. Study of dynamic stability of simply supported beams on Pasternak foundation considering the effect of first transition foundation parameter. Journal of Applied Mathematics and Mechanics (ZAMM). 2012;**92**(6): 490-496. DOI: 10.1002/ zamm.201100040

[15] Subbaratnam B, Jena SP, Reddy DVR. Dynamic stability behavior of simply supported beam resting on elastic foundation and the effect of higher transition foundation on dynamic stability boundaries. In: 13th International Conference Proceedings on Vibration Problems (ICOVP2017). Guwahati, India:

Indian Institute of Technology; November 29 – December 2, 2017

[16] Songsuwan W, Pimsarn M, Wattanasakulpong N. Dynamic responses of functionally graded Sandwich beams resting on elastic foundation under harmonic moving loads. International Journal of Structural Stability and Dynamics. 2018;**18**(09): 1850112

[17] Ying ZG, Ni YQ, Fan L. Parametrically excited stability of periodically supported beams under longitudinal harmonic excitations. International Journal of Structural Stability and Dynamics. 2019;**19**(9): 1950095. DOI: 10.1142/S0219455419500950

[18] Subbaratnam B, Jalagam RR, Jena SP, Gupta MS. Effect of higher transition foundation on dynamic stability boundaries of simply supported beam. IOP Conference Series: Material Science and Engineering. 2020;**998**:1-10. DOI: 10.1088/1757-899X/998/1/012043

[19] Chao X, Wang Z, Li B. Dynamic stability of simply supported beams with multi-harmonic parametric excitation. International Journal of Structural Stability and Dynamics. 2021;**21**(02): 2150027. DOI: 10.1142/S0219455421500279

[20] Deng J, Shahroudi M, Liu K. Dynamic stability and responses of beams on elastic foundations under a parametric load. International Journal of Structural Stability and Dynamics. 2023; **23**(2):2350018. DOI: 10.1142/S0219455423500189

[21] Huang Y, Huang R, Zhang J. Dynamic stability of Nanobeams based on the Reddy's beam theory. Materials. 2023;**16**(4):1626. DOI: 10.3390/ma16041626

Chapter 3

Integrity Assessment for Drill Shafts Foundation in Public and Private Works with Available Technologies in the Twenty-First Century

P. Francisco Centeno

Abstract

The evaluation of the concrete integrity in large-capacity piles (drilled shafts) associated with engineering works, designed to receive service loads from public or private structures, represents an especially important responsibility that is shared among designers, construction engineers, and inspectors. Additionally, this responsibility also lies in the hands of the promoters who may not choose to conduct the investigation for the integrity of the concrete cast *in situ* with the non-destructive tests available at the presentation date of this chapter. Integrity control tests allow one to know in more detail about the continuity of the concrete pouring in each analyzed element and promptly alert the engineers and managers involved in the site about the existence of anomalies due to flaws or construction defects. This information is available with different types of graphics, data, and tomography images that guide us on the continuity of the emptied concrete in the whole body of the piles. The methodologies on display make it easier to search anomalies in 2 dimensions (2D) and 3 dimensions (3D) for possible flaws and defects that may occur in the cap, shaft and the tip of the drilled shaft. These are techniques that allow the structural and geotechnical engineers to visualize what was previously invisible to the eyes and uncertain to everyone's knowledge.

Keywords: crosshole, ultrasonic, monitor, PIT, PET, CHUM, CSL, TIP, piles, tomography, concrete, integrity

1. Introduction

The objective of this article is to present information on three of the nondestructive instrumentation tests, available at the beginning of the twenty-first century, to safely evaluate the integrity of the casting in piles and in large-capacity piles. These tests protect the lives of the users in the structures, where they are used and preserve financial investments used for the construction of public and private works. The use of these tests mitigates the effects of differential settlements, reinforcement corrosion, and functional failures that may occur in the future due to anomalies in the quality of the poured concrete.

In the case of public works, simple or mixed bridges (vehicular and/or with built-in Metro-type mass transit systems) are among the largest, most expensive, and most complex structures that currently exist, which makes them a crucial transportation asset invaluable for the local and regional connectivity of many countries. When the bridges fail, the results can quickly become a human and economic catastrophe for any country due to the isolation and interruption of the movement of people and the different services for their development.

According to international standards, all large-capacity pile or pile foundations involved in vehicular or mass transportation bridges, docks, wind turbines, buildings for multifamily housing, offices, shopping centers, land retention systems for excavations of more than three basements, discontinuous stabilization systems, stations for mass transportation systems, and piles for transmission towers should be evaluated with the integrity tests available. Failing to evaluate the foundation elements represents an extremely considerable risk for these works.

When a single ("mother") pile is built with simple (type 1) or coaxial (type 2) reinforcements (which do not have redundancy of additional elements in their caps), and the execution times required for works (public or private) are accelerated to achieve the goals, this situation can lead to anomalies if they are not reviewed properly due to failures and/or defects in any of the construction stages of the foundations. Any variation in the materials and/or construction processes affects the final integrity of the piers and piles.

With the support of the integrity tests available to date, it is possible to verify and adapt them after their construction with the necessary corrective measures in the event of anomalies due to defects or failures in the head, shaft, or tip from the pile. So far in the twenty-first century, modernized nondestructive testing equipment and increasingly complete calculation programs have been developed making it possible to facilitate the interpretation of the results based on the experiences accumulated by the different specialists who use these tests in different continents.

This article presents the tests based on current ASTM standards, as well as the terminology and evaluation criteria established by the committee of experts that make up the institute of deep foundations (Sellountou, Amir, and others from the DFI – Deep Foundations Institute) [1].

Some experiences collected by Centeno Pulido and Centeno Werner; In Refs. [2, 3] are included in multiple trials carried out with two of these technologies available for public projects (several kilometers of bridge sections, 8 stations on Line 2 of the Panama Metro), industrial buildings, public and private buildings in the city of Caracas, and on the Gamboa bridge of the interoceanic railway (Balboa-Pacific/Colón-Atlantic) in the Panama Canal zone.

In the face of earthquake threats, integrity tests allow geotechnical and structural engineers to mitigate the risk of functional failure due to the presence of partial discontinuities (faults or defects) in piles. They specify the location of uncoated areas to avoid reinforcement corrosion and evaluate the condition of the pile in the face of overloads, lateral loads, wind loads, earthquake loads, and variations in the admissible capacity that may occur in infrastructure elements due to combinations of regulatory loads.

These tests make it possible to evaluate the quality and integrity of the piles poured with concrete (drilled shaft) according to the construction procedures used by the different contractor companies for public and private works.

Before all nondestructive tests that currently exist (CHUM, CSL, PIT, and TIP) were available in Venezuela, only the Testcrete equipment, developed by the Gamma Test company in Israel under Gamma-Gamma technology, was used with frequency Preiss K [4]. This nondestructive test was widely used by the engineers Gustavo Pérez Guerra, José Bernardo Pérez Guerra, Roberto Centeno Werner, Carlos Rodríguez Álvarez, and Mr. Maxim Silberg, all of whom are members of a pioneering group in the use of this technology in Caracas, Venezuela. They formed an expert work and research team that conducted the auscultation of the important project piles during the 1980s.

These tests also allowed improvements in the techniques available for safely excavating and pouring piles using increasingly better equipment from piling companies. They also made it possible to timely repair some of the foundations that were compromised with failures and/or construction defects in diverse types of soil profiles and decomposed rocks, dry or under the groundwater level.

At the time, the Gamma-Gamma test (Testcrete) made it possible to verify the integrity of the excavated and cast piles (cast *in situ*) by means of measuring the concrete density profile with respect to depth. It was also able to demonstrate the importance of reviewing the time when foundations were built for important works.

Minor or major inclusions of soils and/or bentonite bubbles in the concrete mass were promptly detected. These detections frequently occurred in the head (higher % due to concrete contaminated with bentonite or sediments), tip (less than the head and caused by wall collapse), and shaft (the lowest % of the 3 cases).

A historical case to determine the integrity of the foundations was made in an important banking tower in Venezuela. The Gamma-Gamma test was used at the time and allowed the early detection of the existence of a bentonite bubble trapped in the shaft of the pier, which covered an important part of its section. The defect was detected at a depth of −11 meters from the surface.

At first, the piling company questioned the existence of the instrumented anomaly, arguing its extensive experience in the execution of piles and denying the certainty of what the performed test indicated. After overcoming the uncomfortable situation, all of the logistics were organized by the inspection management, and the ground around the pile was excavated with the help of metal shoring to safely reach the affected site.

The size of the defect was determined by the test and caused great concern to the promoter and owner of the work, who supported all the investigations conducted by the inspection and requested to review all the piers to be built in the work with these tests. The repair of the defect made it possible to adapt the foundation. Not having conducted these control tests would have generated long-term structural pathologies in the building due to differential settlements and corrosion of the uncoated reinforcements.

It was concluded that it was always necessary to leave the reinforcement hanging 1 m from the bottom of the excavation to avoid the "bullet point" effect caused by soil falling to the bottom and difficulty in lateral exit of the concrete. Another condition that was verified with this test was the contamination that occurs at the top of the pile in the first 50 cm of the concrete with bentonite when it rises and reaches the surface.

Given the difficulty of managing import permits for the radioactive source (Cesium 137) and the costs associated with this test, it was decided to seek new nondestructive testing technologies that do not use radioactive sources.

The Testcrete test made it possible to demonstrate the importance of being able to safely review the foundations on time, clearing up doubts about the quality of the castings, and creating the confidence needed to continue with the construction of the superstructure. 38 years have elapsed from 1985 to the date of the drafting of this article.

Modern technologies continue to be developed globally to assess the integrity of piles. The PET integrity test, the cross hole ultrasonic monitor (CHUM), and the thermal integrity profiling (TIP) test.

Research has been conducted in Florida, USA to measure concrete coatings through correlations with temperatures (°F) using cable and temperature sensors, as well as in other countries. Research has also been done in Europe using °C with the use of optical fiber distributed in the armor of a circular pile (London, England, 2017). The piles were instrumented in parallel with other instruments such as extensometers, strain gauges, Osterberg cell load test, and CHUM tests based on research papers by Gray Mullins et al. [5] and, Yi Rui et al. [6], respectively.

Between 2004 and 2020, the automation of equipment based on most of these techniques (CHUM, PET, and TIP) has made significant progress, making it easier for inspecting engineers to use it in the field and helping with rapid data processing for timely issuance of quality control reports in the field and in the office. The techniques have matured enough, and there is a truly clear protocol established by the DFI [1].

Currently, there are computer programs developed by means of algorithms that allow the presentation of numerical and graphic results for better interpretations of the anomalies due to failures and defects. The most up-to-date software includes the DFI-approved protocol.

There are engineering works, where piles or piers are used in complicated geological-geotechnical environments. In some instances, as is the case of the bridges for the transport of automotive traffic and mass transport trains, there are some special foundations that are located on rivers that are exposed to the processes of progressive soil erosion (scour foundations) due the effect of the floods of currents and to the varied angles of attack of the water flow on them. Further, information on this topic can be found in papers presented by Eng. Marco Falcon Ascanio and Eng. Mauro Nalesso, Institute of Fluid Mechanics of the UCV, Caracas, Venezuela [7] and Magdi M. from Sudan Africa [8].

Hydraulic erosion by scour often exposes the shaft of piers and piles and leaves them partially uncovered with the risk of partial loss of soil around the area that partially contributes positive lateral friction and determines an important part of the capacity of total load considered in the original design. When turbulent flows occur, incipient corrosion and deterioration processes can also be generated in the event of defect anomalies (discontinuities) due to impacts with rounded pebbles on the head and on the upper shaft of the pile.

Many pile works are subjected to load conditions due to lateral and shear forces. They require careful design and construction to provide sufficient flexural stiffness when required. They must offer sufficient metallic reinforcement and be protected with an adequate coating (concrete) to be able to resist shear stresses, axial load, and moments during their useful life.

Lateral load conditions are frequent in bridges with the same or different diameters, in stabilization walls with discontinuous pile systems (contiguous, tangent, and secant), and in buildings located on fills with variable thickness close to slopes with significant heights. Some considerations on this type of solution with secant piles were presented at the XI Conference Gustavo Pérez Guerra, SVDG 2010, published by Engineer Nelson Rodríguez Delgado [9].

The most recent earthquake-resistant engineering codes, developed in California USA, through research carried out by the University of San Diego in conjunction with the California Department of Transportation (Caltrans) “Seismic Design Criteria” and the U.S Department of Transportation — Federal Highway Administration FHWA-AASHTO use the Load and Resistance Factor Design (LRFD) methodology exposed in several editions and complemented in its latest available edition No.9; May 2020 (mandatory use in the USA and in other countries since 2010).Publication made by Juan Murcia et al. [10], Hassan Muhammad [11], V.R. Sturrup ACI-SP-82 [12] and WaDOT-USA [13].

The LRFD method is based on three states: 1. Service limit state, 2. Stress and fracture limit state and, 3. Extreme event limit state (earthquake, winds, flooding, boat, or vehicle impacts). There are combinations of permanent and transitory loads for each state.

The AASHTO 2020 LRFD methodology allows the design of large-capacity foundation piles (drilled shaft), taking into account full-scale dynamic tests carried out in the laboratory, with loads from beams and columns supported on two types of piles. It is possible to evaluate the behavior under conditions of dynamic loads due to earthquakes for this type of structure.

LRFD works with the service limit state. It allows us to consider the forces, moments, and displacements that govern the design through the resistance of the foundations and the superstructure of bridges. It also considers the allowable stress design for the diverse types of foundations (see: Mn DOT LRFD [14], State of California DOT - Caltrans [15]).

The new AASHTO-LRFD design manual (May 2020) [16] combines ACI-318-19; [17] with Geotechnics. It presents complementary considerations that ACI-318-19 does not do separately since this regulation is only of a structural nature. This information can be found in the course presented by the National Highway Institute (NHI) in April 2007 (LRFD) [18].

2. Integrity testing by ECO (PET)

The Pile Echo Tester (PET) or echo integrity test, also known as the “sonic”, “low strain integrity test”, or pile integrity test (PIT), is a “nondestructive” test that is carried out in accordance with the ASTM-D5882 [19] standard and consists of applying a series of impacts repetitive at the top of the pile by means of a light hammer equipped with a plastic/nylon tip.

The pile must be topless and clean in the horizontal plane without soil, dust, grease, or debris to cushion the impact of the hammer so that the compression wave can develop well at the time of testing. The signal is recorded by means of an accelerometer that is activated with each impact of the hammer on the head and receives the signal back after traveling the entire pile from the head to the tip (round trip).

It is particularly important that the accelerometer is well adhered to the concrete surface of the head. The site to be evaluated must be the center of the pile and separated by at least 6″ from where the blows are made. Good adherence of the accelerometer with the putty must be guaranteed to allow intimate contact between the accelerometer and the clean concrete (tip). Only in this way can the wave be recorded well throughout its round-trip trajectory. Several sites of the pile section can be analyzed, avoiding coincidence with the main reinforcement. It is a fast and cheap test and is the most widely used in piling works around the world. Due to its low cost, it is an extremely popular test among builders and inspection engineers.

The result is an echo reflectogram that presents the impedance of the compression wave with respect to depth. From the test, the speed of the wave is obtained by means of the quotient between the detected length and the total round trip time divided by 2. The speed can be correlated with the quality of the poured concrete. The test determines the location of the pile toe in rock or hard ground with good accuracy, except in some exceptional cases, and allows us to qualitatively rule out whether there is any significant interruption of the concrete in the shaft of each pile.

In the case of driven piles, the position of the joints (in elevation) between the segments of the driven pile must be well known so as not to confuse the received signal when it passes through a joint since it would present it as if it were a defect (impedance change in the reflectogram).

When there is an anomaly due to failure or defect in the pile shaft (generated by some inclusion of major earth, air, or bentonite), the round trip of the wave is modified in the echo response of the compression wave. The site where the anomaly occurs can be determined, but not its size in the sectional area or the shape of the volume where it occurs.

There are many types of reflectograms or "echo curves" that can be recovered on a job site during PET testing, depending on the type of pile, the concrete used, the location of the rock, and the rate of setting time. A gallery of reflectogram types with their analysis is detailed in work published by Webster et al., [20] and by Liqun Liang and Frank Rausche; [21].

The shaft of the pile generates friction on the walls (skin friction) to the wave during its round trip to the accelerometer. As a result, the reflectogram has a limitation that depends on the slenderness (ratio between length and diameter) and the types of strata.

It is important to carefully review the sign convention used in the equipment to be able to interpret the start and arrival of the compression wave ("X" axis) in order to fully understand whether it is an anomaly due to excess (over volume) or defect (empty) with respect to the "Y" axis (depth). See **Figures 1** and **2**.

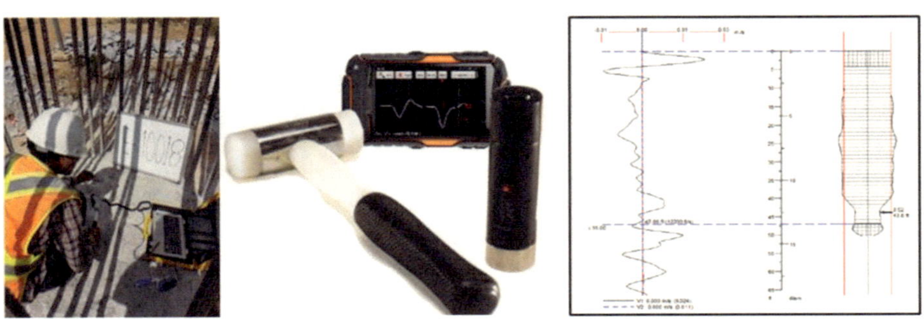

Figure 1.
PET/PIT test with reflectogram graphic by PDI (Ohio – USA) or pile test (England) in a tested pile. It shows an interpretation of how the bottom of the pile was left. A bottleneck anomaly (necking) at the tip of the pile is represented. Accelerometer and hammer. A cellular phone or laptop is needed to register the data.

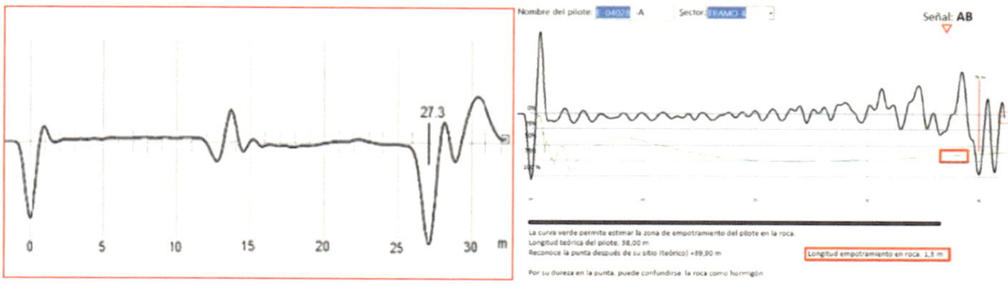

Figure 2.
PET (record low strain) on a driven pile evaluated. The jump is observed right in the middle (it coincides with the joint between 12 m precast segments of the pile). Left; [22]. Presented by Amir Joram from Piletest - Israel. Right reflectogram with impedance measured by Centeno [2] on an excavated and cast-in-place pile. Mistakes in the pile tip below the theoretical length [twenty]. It also shows the Fourier transform curve in green (heterogeneous pulse soils and in rock (asymptotic flat area).

3. Crosshole ultrasonic monitor (CHUM) test

The crosshole ultrasonic monitor (CHUM) or crosshole sonic logging (CSL) is nondestructive and is governed by the ASTM D-6760-14 standard, [23]. It is also known as a sonic transparency test.

The frequency emitting equipment must work between 50 kHz and 60 kHz. This is the best frequency available to achieve conclusive and proportional results to the investment and cost of the equipment. There are other equipment that work with higher frequencies between 100 kHz and 250 kHz, but larger drives are required and are not very profitable due to their price, in addition to needing to handle exceptionally large files that are uncomfortable to interpret.

The two most important parameters that are measured are the FAT (First Arrival Time - red curve), which corresponds to the first arrival wave of the ultrasonic pulse from the emitting hydrophone to the receiving hydrophone (time in milliseconds), and the energy reduction (green curve) that is measured in decibels (dB) see **Figure 3** prepared by Joram Amir [22, 24] from Piletest (**Figure 4**).

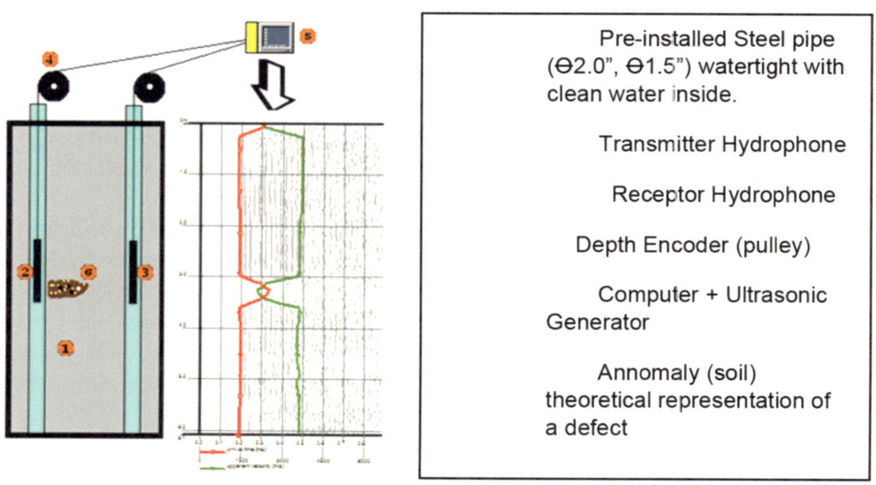

Figure 3.
Cross hole ultrasonic monitor (CHUM). Source: Amir Joram [22, 24].

Figure 4.
Equipment for cross hole ultrasonic monitor (CHUM) test. Ø 225 cm diameter Centeno-Rodríguez y Asociados s.a. Panamá metro line 2 from 2016; [3]. Pile excavation and the concrete bridge are in operation.

Defining the faults and/or defects in a pile is a process that requires making several measurements with the hydrophones from the bottom of the pile to the surface. Combinations of readings are made between the pipes installed in all directions. It requires experience and dedication to be able to correctly interpret the obtained results.

Opposite diagonals (the most important) and chords (complement) that are formed between the different configurations and combinations from one tube are measured, with the remaining group of tubes installed. All the tubes go inside the reinforcement of the pile before concrete pouring.

The number of pipes that must be installed in the pile is estimated as a whole number that results from dividing the diameter of the pile (cm) by 30 cm. For example, if we have a pile of 215 cm internal diameter in the reinforcement or 225 cm external diameter (excavation), a minimum of seven tubes is required, with 21 combinations of readings between the seven tubes. Joram and Erez Amir, Israel [25] and [22]. Placing fewer control tubes causes blind areas in the evaluation due to lack of access. This has been verified in multiple works around the world (**Figure 5**).

The tubes should be made of steel, preferably 3 mm thick and with a minimum diameter of 1.5″ or up to a maximum of 2″, to facilitate the passage of the hydrophones. The tubes must be joined with the help of threaded sleeves (male–female), or rubber joints with clamps, that allow the coupling of the segments of steel tubes available in the market. The most important thing is that the pipe has well adhered to the cage with the wire and that the joints guarantee watertightness without water leakage.

The pipes are not placed outside the reinforcement (cage) because they can hinder the process of lowering the cage into the excavation and can bend, deviate, and/or open between their joints. PVC pipes should not be used in piers or piles because they tend to debond after a few days during the hardening of the concrete (high setting

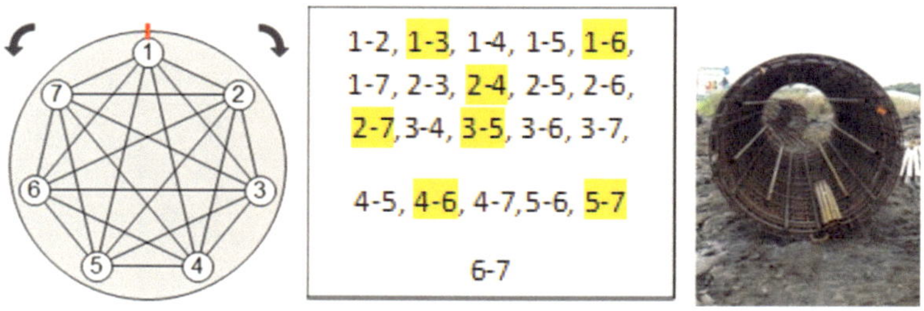

Figure 5.
Number of sonic tubes and all combinations.

temperatures) and produce a separation (air ring) through which the ultrasonic waves escape, altering the test results (FAT and RE).

It is especially important to avoid the use of grease at the sleeve joints with each run of threaded pipe because this alters the travel of the ultrasonic waves, and the hydrophone can become contaminated with the grease. The joining between armature bodies is an activity that requires extreme care to be able to join the cages very well and connect all the sonic tubes together. Good practice leads to conducting a precise installation work that avoids deviations and openings of the joints between the tubes. Preferably, male–female steel tubes (quick coupling) or flexible rubber sleeves with their clamps should be used. Threaded sleeves are more difficult to join between two cages when they are being lowered and take time out of the construction process. Security is required during this activity to avoid accidents with the hands of the workers.

When the (FAT) increases and the (RE) decreases in the same place of the pile, it is due to the existence of an anomaly. Thresholds are then determined to conclude if it is a failure (contamination of the concrete) or a defect (discontinuity due to a hole or strangulation with the earth). Below is a photo showing an example of a defect detected with the CHUM test, located at the head of a bridge foundation pile. It is a precise test that checks the defect of the concrete that traveled from the bottom to the top on the surface (poor or contaminated concrete area). We proceeded to scarify, clean well, and replace with high-resistance fluid mortar and epoxy anchor.

Figure 6 shows a typical defect due to interrupted emptying (voids) [3, 24]. The head of the pile was completely scarified and the contaminated concrete was removed in a thickness of 50 cm until reaching the sound concrete. This activity is frequently conducted in the construction of piles. The scarification thicknesses are proportional

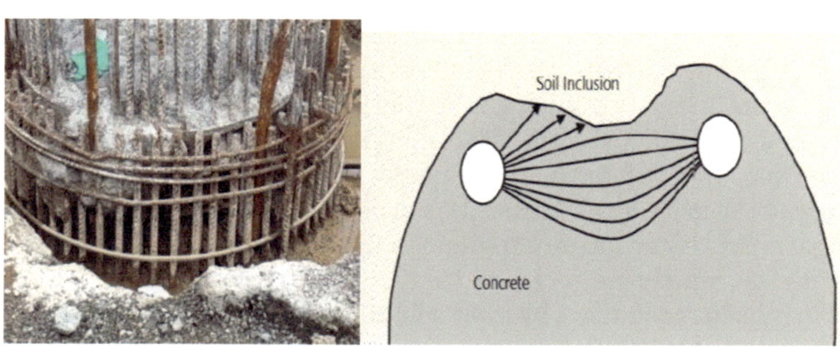

Figure 6.
Example of the void and soil inclusion in the head.

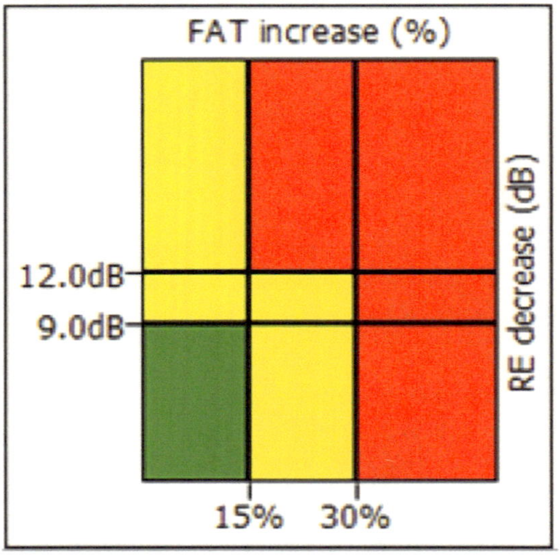

Figure 7.
Source deep foundation institute (DFI). October 2019 [1]. Green: CLASS A (ACCEPTABLE); Yellow: CLASS B (CONDITIONALLY ACCEPTABLE); Red: CLASS C (ANOMALY); Fault Anomaly: (15–30%) FAT RE (9–12 dB) 50% or more combinations at the same depth; Default Anomaly: (>30%) FAT RE > (12 dB) 2 combinations minimum at the same depth.

to the length of the pile and to the existence of concrete contaminated in the upper part with soil or bentonite. The longer the pile is, the scarification thickness on the head may increase (**Figure 7**).

Previously, cascade-type (waterfall)images were used to represent the graphic output of the test. Currently, the FAT (first wave arrival time) and RE (energy reduction) curves are also represented in the cascade-type base. Other options also show the previous two, including speed curves (French standard) and PSD curves (Chinese standard). The software we use for research is updated to July 2020 and considers the protocol of the deep foundations institute [1] and Carlos Fernández Tadeo [26]. Centeno-Francisco uses this most recent methodology and highlights his experience gained during multiple ultrasound integrity tests conducted in various works in Panama [3].

Below are three examples evaluated by the author of this article showing the results of typical graphs (waterfall) of the CHUM assay. A case of a healthy pile (left) and two cases with anomalies due to failure and defects in the shaft and tip of a Ø2.25 m pile are presented (**Figures 8–10**).

It is possible to determine the position of a defect or a failure in the concrete of the pile with the CHUM test. When it is possible to safely excavate to the depth where the defect or failure anomaly occurs, it is possible to verify that it is in the place where the test indicated it. Tipping is done before digging. Arriving at the site, the affected area is cleaned very well and repaired in a timely manner, avoiding future damage to the infrastructure of the project. This procedure avoids altering the admissible load capacity and mitigates the risk of future corrosion in the reinforcement steel bars of the piles in contact with soil, weathered rocks, and water (if it had not had an adequate coating).

This test has been recognized by most pile companies, owners, and project managers for many years. Initially, some field operators thought that their results were inconclusive and that they delayed production. Today, serious piling companies recognize its effectiveness and support its use since it allows correcting defects that

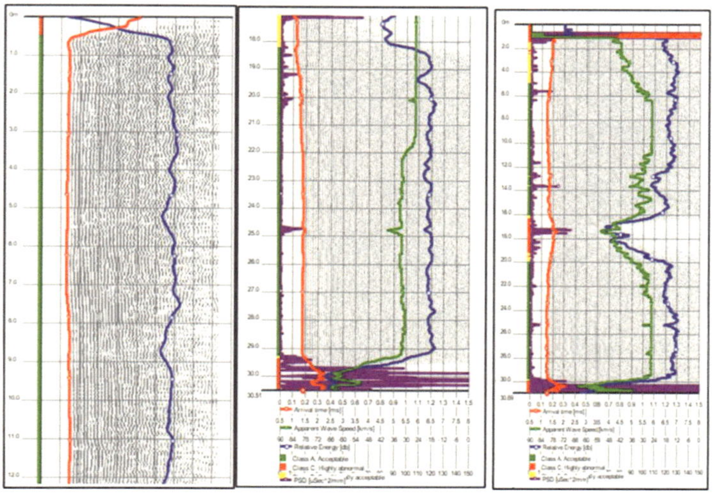

Figure 8.
Good (L), shaft defect (pile tip) (C), and defect and a dirty tip (R) [3].

Figure 9.
CHUM tests for the Panama Canal interoceanic railway bridge. LRFD (load and resistance factor design) AASHTO. Piles in the top head were subjected to daily cyclical loads due to the passage of the locomotive and many wagons (2 floors with 40 pc containers each) [3] September 2020. Centeno-Rodríguez y Asociados s.a.

Figure 10.
Example of 2D and 3D tomography (longitudinal profile). Two photos with examples of anomalies detected in the head and shaft of two piles using the CHUM test (cross hole ultrasonic monitor). The author of this article is photographed beside an especially important anomaly detected in the shaft of this pile (−5.00 m depth and 1.5 m height)). No concrete for steel bars (covering) [3].

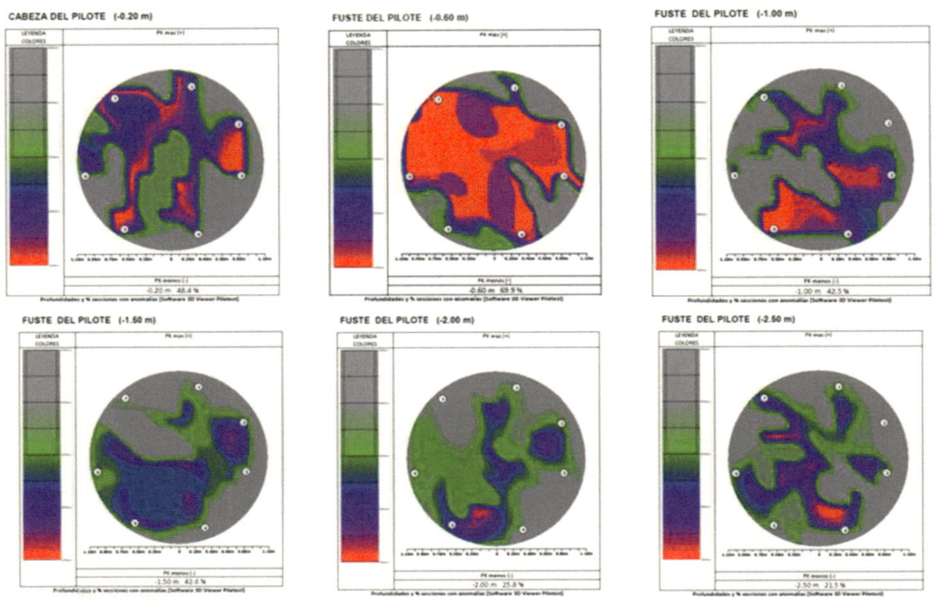

Figure 11.
Some digraphs or slices evaluated between −0.20 m and – 2.50 m are presented, indicating the distribution of % anomalies through images with speeds less than 3000 m/sec. Red color indicates water and a probable vacuum. Cuts are presented in various places, where the anomaly is recorded. Gray color (excellent), green (good - limit), blue (poor), and red (void). Tomography analysis conducted by the author of this article using **Piletest ®** *special tomography software (licensed). Israel-England.*

may have occurred during construction. Future legal problems between the parties are avoided (**Figure 11**).

4. Limitations of the CHUM test

1. The main limitation is that it does not allow the integrity of the concrete to be measured in the area of the thickness of the external coating to the steel because the sonic tubes are inside the reinforcement (it does not cover the extrados of the reinforcement – 7 to 10 cm outside the steel cage).

2. A minimum of 7 days is required for the concrete to set and harden. The setting temperature of the concrete must be low enough (< 65°C) so as not to damage the hydrophones, and thus be able to operate without damaging the internal electronics. Measuring within less than 5 days could present false positives in the signals. From experience, the FAT and reduction energy values would change if the CHUM test were conducted again, after more than 15 days. The idea is to work between 7 and 10 days for the first reading.

3. When drilling with core-drill equipment for repairs and grout injections in the affected area, there is previously a healthy hard concrete in the unaffected areas and located before reaching the site with the defect. Placing a grout and/or mortar under pressure, still setting after 7 to 10 days in the removed healthy area, will present less resistance than the original concrete placed because the grout

does not normally contain the coarse aggregate. The CHUM signal could then present some FAT and reduction of energy (RE) variations that show a partial improvement and not 100%, where the original defect/failure detected with the test occurred (heterogeneous mixture of concrete and grout).

4. It is only possible to measure at the tip of the pile up to the height, where the sonic tube has been left in the reinforcement. Lowering the tube further can buckle the sonic tubing if it rests on the tip. It is not possible to measure the area without the tube. If the reinforcement is lowered further to the bottom, the concrete will not fill the tip well and a "bullet point" effect can be generated in the pile. Normally, the bottom armor is left hanging between 20 cm and 50 cm in many cases.

5. Integrity test with thermal profile (TIP)

This integrity test is the newest of the three tests presented in this article and is known as thermal integrity profiling (TIP). It is governed by the ASTM D-7949-14 standard. Ref. [27]. It was developed at the University of South Florida by Dr. Gray Mullins and others. Ref [5] (other useful references: Ryan Allin [28], Gina Beim [29]). For 16 years, important advances have been made with this methodology, and its use has become popular. It is used to control piles on public transportation bridges in the state of Florida (Clearwater-Tampa) and several other US states, as well as in some other countries.

The TIP test is another reliable alternative for the control of concrete integrity in piles and piles excavated and cast in place (drilled shaft - cast in place). It is also frequently used to assess the quality of the pouring in micropiles, secant pile screens, discontinuous screens, and jet grouting columns, among others.

The test consists of making a continuous record of the setting temperatures of the cement throughout the entire pile while the curing and hardening of the concrete occurs in the first 48 hours.

The values of all temperatures that develop in the concrete mass during curing are recorded. Temperatures are recorded by strategically located thermal sensors distributed throughout the armor (cage). The tubes and sensors are distributed in the same way in the section as in the CHUM or CSL test; the number of vertical lines of sensors is obtained by dividing the diameter of the reinforcement (cm) by 30 cm. The approximate whole number of the resulting quotient is always used.

For piles of 2.15 m internal diameter of the cage or 2.25 m diameter excavation, seven vertical lines are used and proportionally separated from each other on the perimeter (215 cm/30 cm = 7). For 1.20 m piles, four vertical lines (120 cm/30 cm = 4) are used. The tubes can be made of PVC or steel and with the same spacing characteristics as in the CHUM test, in case of using a sensor that travels through the tube. There exists the alternative of leaving expendable sensors (they are consumed (fungible) and cannot be recovered) with their use inside the poured concrete of the pile.

Regarding the "debonding" effect of PVC with concrete, this material (PCV) does not affect the results. It also does not require the tubes to be filled with water to perform the test. The temperature that is measured is the one that is registered at each advance height during the setting of the cement inside the tube. If the tubes are made of steel, they can also be verified with the CHUM test after 7 days of casting (**Figure 12**).

Figure 12.
Distribution of 10 tubes per stack (3.00 m) TIP test [30]. Execution of the test and recording sensor used [5]. Photos: Work conducted by PDI - Ohio USA.

Lower temperatures are recorded near the surface and higher temperatures are recorded as you go deeper into the stem. On the right, the temperature in different sectors is presented. It was determined that the reinforcement was slightly out of phase or displaced with respect to the perforation with the help of other tubes with little coating. In **Figure 13**, it is easy to visualize higher temperatures in areas with greater coverage on the left (red point) and lower temperatures (blue point). On the right there is another image that shows the longitudinal section of a pile, indicating the profile with the distribution of temperatures of the pile between the surface and the toe. Gray Mullins [5].

The colder regions can be associated with inclusions of soil or hollow zones without concrete (colder). In the upper part (head) and in the bottom (tip), the temperatures are also lower because they are in contact with air above and with earth or rock in contact with the area near the tip.

When the temperatures recorded during setting are above the average values in the diverse types of concrete, these may be related to an excess volume of the concrete mass (bulges) associated with hotter areas due to having more cement. At the tip of the pile and at the top of the pile, it has been determined that there are two zones that always present important changes in temperature (colder zones) that generally correspond to a height proportional to a diameter or a length equal to 40% of the square root of the temperature measured in degrees Fahrenheit. They are affected by their proximity to the ground and/or rock (at the tip) and to the atmosphere (head).

Two important concepts are managed in the test during cement curing (concrete setting) according to research conducted at USF - Fl, USA by Professor Gray Mullins and others [5]:

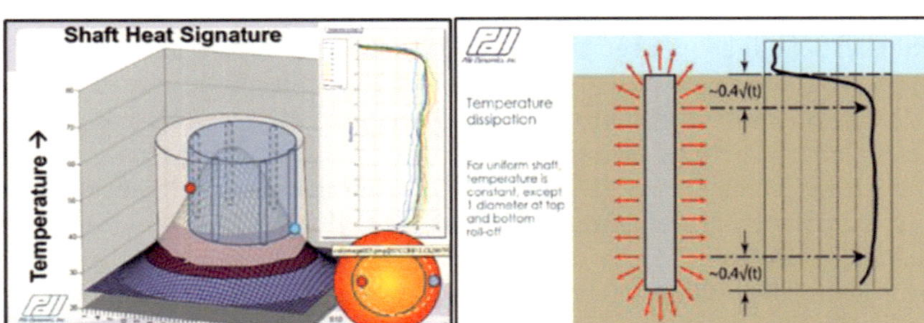

Figure 13.
Distribution of temperatures in the concrete of a pile with respect to the ends (head and tip). A pyramidal temperature distribution is observed under the pile head and tip (Gray Mullins) [5].

1. *Heat production*, which represents the amount of heat produced during the exothermic reaction of cement curing and depends on the amounts of materials and the type of cement used in the mix design (cement, water, sand, and stones) for used concrete.

2. *Heat diffusion*, which is just as important as the heat production that occurs around the pile in the different soil and rock strata. It also depends on the position of the groundwater level in the subsoil.

The flow of heat that is generated and diffused involves simultaneous mechanisms of conduction, convection, and radiation. Conduction is the mechanism that dominates the transport of heat. The heat capacity in soils is based on the volumetric fraction of the three phases: solid, voids (gas), and water.

There are two parameters that affect the dissipation of heat from the concrete mass to the soils: the dissipation of heat around the pile and the resistance offered by the layers in being heated. The denser the soil and hard rock around the pile, the easier the conductivity of heat in the soil or rock; they require more energy for these to be heated. Hence, there are few hours available for the heat to be maintained and transmitted efficiently during its production. It is important to take measurements during the first 48 hours after pouring.

The TIP test also allows detection if the reinforcement has been well supported or deformed against any of the walls of the excavation and planting. It also detects the presence of failures and defects in any part of the pile. It allows the evaluation of the perimeter area of the concrete pouring (covering) outside the reinforcement. Regarding the CHUM-CSL method (only effective in the internal zone of the reinforcement), TIP does allow us to be able to evaluate a little more of the extrados of the tubes with more precision. The Gamma-Gamma test also allows a little more evaluation of the thickness of the coating on the back of the reinforcement based on the density of the concrete, but less than TIP (**Figures 14** and **15**).

The TIP test allows the engineer to estimate the shape of the shell (radius profile) with respect to the depth. From the data and with the software developed by the USF and PDI, the resulting geometry from the pouring of concrete in the pile or pile can be graphed using computer algorithms. They are image generators based on the integration of data recorded in all the tubes and temperature curves during the first 48 hours of setting.

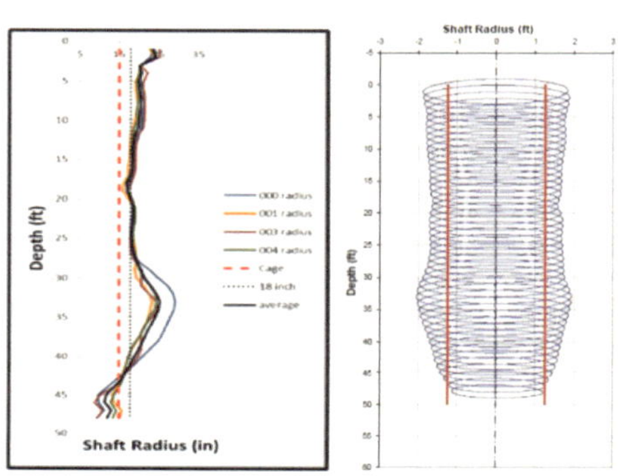

Figure 14.
*Determination of the pile geometry (radius profile (inches) from the TIP test by correlation with the temperatures (°F) measured in a sound pile. Image "rendering" [13]. Correlation: Pile radius (inches) = [(0.81*T(°F)) - 51.3]. (average temperatures). WaDOT. Mullins Gray.*

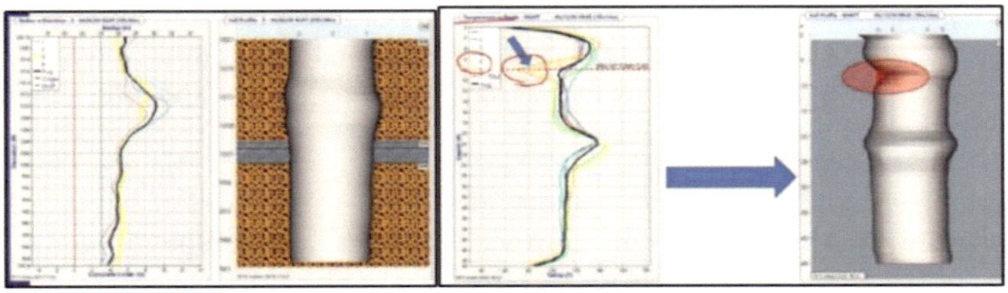

Figure 15.
Examples of TIP curves with temperature dissipation in two piles, one showing a bulge (L) and one showing a local defect at the top (R). Reference: PDI Webinar; OH.USA; June (2020) by Ryan C. Allin [28].

Automated sensors that automatically record all setting temperatures can be used with a datalogger (TAG) and a recorder per tube (TAP). They can then store the data in the cloud and be consulted remotely from anywhere in the world or from the office at any time. PDI technology available.

The recorded temperature depends fundamentally on the type of concrete used in the mix design, the cement content, diameter, length of the pile or pile, and setting time (hours). The larger the diameter of the pile and the length, the higher the setting temperature of the cement in it will be, and it will remain hot for more hours. A graph is presented below to better understand the results.

6. Limitations of the TIP test

1. The optimal time to conduct the test is indicated in the window of **Figure 16**; [20]. It is variable and depends on the diameter of the pile and the type of concrete mix. The smaller the diameter of the pile, the less time is available to collect the data. The greater the diameter of the pile or pier, the more hours of setting are achieved (> volume of concrete and cement content). Reliable data can be

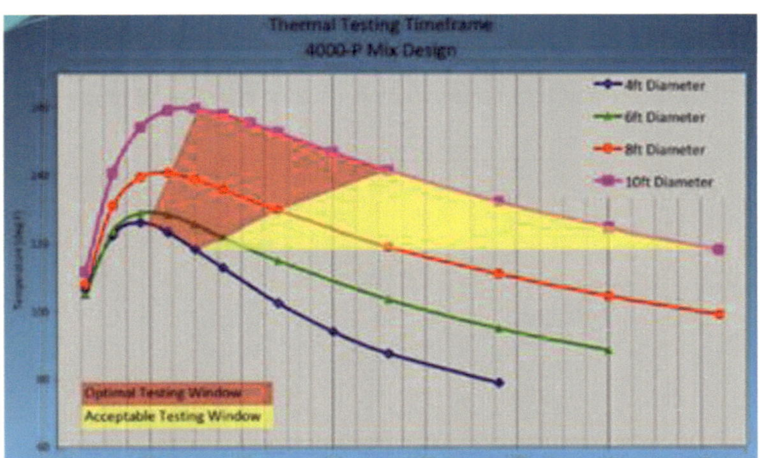

Figure 16.
Different expected temperatures in concrete with mix design: 280 kg/cm2 (4000 psi) 28 Mpa; for different diameters (1.2 m to 3.0 m), temperatures (80°F and 160°F) and setting time (0 h to 145 h) 6 days. Washington Department of Transportation Final Report (2011) [13], page 155. (Brent Robbinson PDI-OH-USA).

measured between two and six days (depending on the size of the pile and pile). The data is less accurate after this time due to the significant decrease in temperatures. It is recommended to automate the measurements as much as possible with the help of the TAG data logger and the TAP devices (sensors) so as not to lose valuable information.

2. The rehearsal should be scheduled in good time. The installation and measurement in the field must be fast and efficient so as not to lose the valid time range to be able to read the temperatures during the entire setting. For piles with diameters of 1.2 m to 1.8 m, it is available from day two to day four of setting to measure. For piles with diameters of 2.0 m to 3.0 m, it is available from day two to day six of setting to measure. Between day seven and day ten, there is still some of the setting heat, but the data is not as precise, see **Figure 16** (x-axis). Reading is not recommended below 120°F (49°C). The threshold can be extended up to 100°F (38°C) in some, but it is not usual. Measurements after 48 hours to verify any injection repair process must be conducted with another test, such as CHUM or CSL.

3. It can be measured with PVC or metal pipes, but they must be empty and without water. It does not matter if there is a punctual detachment or separation of the PVC pipe with the concrete. This condition does not affect the temperature or the results. Water in the tubes prevents the thermal sensor from working well and can reduce the temperature.

4. Correlations between temperature and concrete maturity compressive strength can be made but are not an intrinsic component. They depend on the mixture design and other stochastic variables.

7. Conclusions

Pile integrity tests help the quality manager of a project to make reliable decisions so that their work allows them to lead and have the necessary technical arguments to inform the professionals involved in the design and construction log of the piles. It is used to check whether an adequate or defective construction process has occurred, avoiding negative consequences that cause future performance problems for the work. The quality of the integrity of the piles determines the payment of the elements that are accepted or rejected in a work. Expansion on this subject can also be read with other related articles: Bottiau [31], Camp [30], Grosse [32], and Mark Gaines [33].

The CHUM or CSL integrity test has more than 30 years of development and has been successfully applied on five continents. It is a highly documented test with hundreds of thousands of piles, and piles evaluated in many countries around the world. It is considered a very mature and reliable technique to identify, where the anomalies due to failures and/or defects in piles are located. The TIP integrity test has only been on the market for eight years and is beginning to be documented more frequently.

Controls with the CHUM, CSL, PET, and TIP tests should always be compared with the construction log before rejecting or approving a pile. PET is the most limited of the three tests. Certifying a pile with PET is not recommended for bridges and buildings with large- capacity piers (> Ø1.20 m). Load testing can always be used as a complementary method when a major defect anomaly is observed in a complex and inaccessible position of the shaft or on the tip.

The number of tests required is not defined based on single regulations. Each country has its criteria based on the requirements of structural designers and geotechnical engineers, who will determine the number of piles that must be evaluated on site for each project based on their location, service loads, and redundancy of elements for each column. In mass transportation, works located in seismic zones, where there is only one pile (mother) per column; all piles must be evaluated without exception. In other works, where there is redundancy of piles due to foundations (bridges and buildings), 25% of all piles are evaluated. If more than 20% have defects or failures, the amount is doubled to 25% more. If there are more than 20% with defects or failures, all the piles in the work must be evaluated without exception. In the USA, a greater number of piles are evaluated than in Latin America based on geotechnical and structural criteria. The resistance of any chain is that of the weakest link.

According to the world CHUM test statistics, most defects and failures occur at the head and toe of piles; a lower percentage occurs in the shaft of the piles. Each project has its difficulties, and it will only be possible at the end of the pile and pier work to know in detail how these percentages vary. It is preferable to always work with the FAT (time) and not with the velocity of the ultrasound waves because the tubes are never perfectly aligned and their distance changes with respect to the depth frequently and depends on the installation.

Acknowledgements

To the engineers Roberto Centeno Werner, Carlos Rodríguez Álvarez, and José Bernardo Pérez Guerra for their comments and transmission of previous experiences as pioneers and as users of nondestructive tests (Gamma-Gamma) in Venezuela. To Jack Mulcrone for reviewing this paper.

This chapter is an English translation of the partial original article wrote by the authors: Centeno Pulido F. and Centeno Werner R. "Evaluación de la integridad en pilotes y pilas para obras públicas y privadas con tecnologías disponibles a comienzos del siglo XXI. In Academia Nacional de la Ingeniería y el Hábitat, ANIH Boletín 54 – Memorias del Congreso Venezolano de Geotecnia 2020. Parte 2; 2022. p. 42–56".

Author details

P. Francisco Centeno
Metropolitana University, Caracas, Venezuela

*Address all correspondence to: francisco.centeno@gmail.com

References

[1] Sellountou EA, Amir Joram M, et al. Terminology and Evaluation Criteria of Cross Hole Sonic Logging as Applied to Deep Foundations. Deep Foundations Institute; 2019

[2] Pulido C, Francisco A, Werner C, Roberto R. Integridad del concreto vaciado en Pilas y Pilotes. Revista Cámara Venezolana Construcción. 2015;7(5):36-46

[3] Pulido C, Francisco A, Centeno-Rodriguez Y Asociados. Data bank with non-destructive integrity tests on piles (PET and CHUM) 550 piles: Line 2 of the Panama Metro, and Piles of the Gamboa Bridge (Pacific-Atlantic interoceanic) Panamá Canal Railway Company. Eng. Francisco Centeno Pulido / Tech. Jorge Santamaría. 2016-2020

[4] Preiss K, Caiserman A. Non-destructive Integrity Testing of Bored Piles by Gamma Ray Scattering. Negev, Beer-Sheva, Israel, Ground Engineering: Department of Mechanical Engineering, Ben Gurion University; 1971

[5] Kevin RJ. Advancements in Thermal Integrity Profiling Data Analysis. 2016

[6] Riu Y, Kechavarzi C. Integrity of Pile cover using distributed fibre optic sensing. Sensors Article - MDPI Journal. 2017

[7] Falcon AM, Mauro N. Flow angle of attack at perpendicular bridge crossings. In: Briaud J-L, editor. International Symposium on Scour of Foundations. Melbourne, Australia on November 19, 2000. Miembro Comité TC-33 de la ISSMGE. UCV-Venezuela: Instituto Mecánica de Fluidos; 2000. Available from: https://izw.baw.de/publikationen/tc213/0/scour-of-foundations-symposium-2000-content-pages.pdf

[8] Zumrawi ME, Abusim HA. A Study on Scour Failure at Bridge Foundation. 2019

[9] Nelson RD. XI Conferencia Gustavo Pérez Guerra, "Sostenimientos discontinuos como técnicas de estabilización". XIX Seminario SVDG. 2010

[10] Murcia J, Delso YL. Development length of column reinforcement extended into enlarged pile shafts. In: 7th National Seismic Conference on Bridges & Highways, May 20-22 2013. Oakland, CA; 2013

[11] Muhammad H, Nespak O. Review of Concrete-Filled Steel Tubes (CSFST) "Drilled Shaft Design Requirements" from US DOTs, FHWA, AASHTO and Others. 287 pages. Available from: lNCHRP12-93_Appendices.pdf

[12] Sturrup VR et al. Pulse velocity as a measure of concrete compressive strength. In: In Situ/Non-destructive Testing of Concrete. ACI SP-82. Detroit; 1984. pp. 201-227

[13] Wadot MG, Winters D, et al. Infrared Thermal Integrity Tester. Quality Assurance Test Method to Detect Drilled Shaft Defects. USA: University of South Florida; 2011

[14] MnDOT. LRFD Bridge Design Manual. MN Western, Kevin, State Bridge Engineer: Minnesota Department of Transportation, Oakdale; 2020

[15] U.S. California DOT. D. Engineering Services (Caltrans); "Foundation Manual." 2008

[16] American Association of State Highway and Transportation Officials

(AASHTO). LRFD Bridge Design Specifications. novena ed. USA; 2020

[17] American Concrete Institute (A.C.I.), Building Code Requirements for Structural Concrete (ACI 318-19) Commentary on Building Code Requirements for Structural Concrete (ACI 318R-19). 2019

[18] National Highway Institute (NHI) Course No. 130081, 130081A, and 130081B "Load and Resistance Factor Design (LRFD) for Highway Bridge Superstructures", Publication No. FHWA-NHI-15-047. April 2007, Revised July (2015) 1694 pages

[19] ASTM Standard D-5882. Standard Test Method for Low Strain Impact Integrity Testing of Deep Foundations. West Conshohocken, PA: ASTM International; 2000

[20] Webster K, Rausche F, Webster S. Pile and shaft integrity test results, classification, acceptance and/or rejection. In: TRB 2011 Annual Meeting. USA; 2011

[21] Liang L, Rausche F. Quality Assessment and Procedure Classifications of Cast in Place Shaft Using Low Strain Dynamics Test (LSDT). OH, USA: Pile Test Inc and GRE Engineers Inc.; 2008

[22] Amir Joram M. Cross Hole Ultrasonic Monitor (CHUM) Manual. Herzlia, Israel: Piletest.com Ltd; 2020. p. 77

[23] ASTM Standard D-6760. Standard Test Method for Integrity Testing of Concrete Deep Foundations by Ultrasonic Crosshole Testing. ASTM International; 2014. Available from: www.astm.org

[24] Amir Joram M, Amir Erez I. Relative Energy in Cross Hole Ultrasonic. PileTest Technical Notes. Herzlia, Israel: Piletest.com Ltd; 2016. p. 5

[25] Amir Joram M. Pile Integrity Testing: History, Present Situation and Future Agenda. Herzlia, Israel: Piletest.com Ltd; 2020. 16 pages. Available from: https://www.piletest.com/about-us

[26] Carlos FT. Nuevas recomendaciones Deep Foundations Institute (DFI) en evaluación de resultados de ensayos ultrasónicos cross-hole en pilotes. España: BCN; 2019. Available from: https://fernandeztadeo.com/?p=5027

[27] ASTM Standard D-7949. Standard Test Method for Thermal Integrity Profiling of Concrete Deep Foundations Deep Foundations. West Conshohocken, PA, www.astm.org: ASTM International; 2014

[28] Allin R, Pile Dynamics Inc. Webinar: Case studies. Cost and time savings. In: Achieved through Thermal Integrity Profilling. Cloud Enable. USA: Pile Dynamics Inc (PDI); 2020

[29] Gina B, Robinson B. Current Practices and New Methods. Pile Dinamics, Inc; 2015

[30] Camp WM, Holley DW, Canivan GJ. Crosshole Sonic Logging of South Carolina Drilled Shafts: A Five Year Summary. Denver, CO: New Peaks in Geotechnics; 2007

[31] Bottiau M. Recent advances in pile design, construction, monitoring and testing. In: Proceedings of the XVII ECSMGE (2019) "Geotechnical Engineering Foundation of the Future". 2019

[32] Grosse C. Quality management of fresh concrete - Fresh concrete analysis with ultrasound. In: International Symposium Non-Destructive Testing

(NDTCE). Berlin: GS-NDT; 2003. pp. 26-32 BFT vol. 6, (2005) Berlin, Germany. "New developments in quality control of concrete using ultrasound"

[33] Gaines M. Thermal integrity profiling as a drilled shaft quality assurance tool. In: Western Bridges Engineers Seminars. USA: Washington State Department of Transportation (WaDOT); 2015

Chapter 4

Soil-Structure Interaction: Understanding and Mitigating Challenges

Ali Akbar Firoozi and Ali Asghar Firoozi

Abstract

This chapter provides a comprehensive analysis of Soil-Structure Interaction (SSI), a key component in solving complex engineering challenges amidst rapid urbanization and changing environmental conditions. It elucidates the theoretical principles and practical implications of SSI, emphasizing its role in creating sustainable and resilient engineering solutions. The chapter explores the soil's response to different load scenarios, highlighting the impact on structural reliability and integrity. The narrative includes real-world case studies demonstrating the practical application of SSI principles, advocating their integration into contemporary construction methodologies for improved structural safety. It also outlines innovative strategies to tackle SSI-related challenges, such as employing advanced materials and computational models. Finally, the focus is placed on sustainability and resilience-driven solutions designed to withstand the tests of time and climate change. Serving as a valuable guide for various stakeholders in the field, this chapter underscores the significance of SSI in the development of environmentally conscious and structurally robust constructions.

Keywords: soil-structure interaction, theoretical mechanics, structural stability, mitigation strategies, sustainable practices, construction resilience, geotechnical engineering, impact of climate change

1. Introduction

Soil-structure interaction (SSI) is a crucial concept in geotechnical and earthquake engineering that contemplates the interplay between structures and the ground they rest on. This interaction significantly influences the behavior of structures during events like earthquakes. Broadly, SSI encompasses three primary components:

i. *Structure:* This denotes the constructed building or infrastructure. Factors such as the structure's weight, stiffness, shape, and material properties dictate its interaction with the soil.

ii. *Foundation:* This represents the interface between the structure and the ground, transmitting loads from the structure to the soil. The foundation type (deep, shallow, pile, etc.) and its properties significantly impact the interaction between the structure and the soil.

iii. *Soil:* This refers to the ground on which the structure stands. Properties like soil type (clay, sand, rock, etc.), stiffness, density, and stratification greatly influence its interaction with the structure. Soil can amplify or attenuate seismic waves, altering how an earthquake impacts a structure.

When an earthquake occurs, the ground's motion transfers to the structure through the foundation. Structure responds according to its dynamic characteristics and the nature of the ground movement. Subsequently, structure's motion affects the ground motion, giving rise to a complex interplay known as soil-structure interaction. SSI holds significant implications for the safety, efficacy, and sustainability of structures. It encapsulates the mutual influence between a structure and the soil supporting it [1–3]. Understanding SSI is essential for designing structures resilient to various load types, from static to dynamic loads induced by earthquakes. Technological advances and computational models have equipped engineers with tools to better understand and analyze SSI. However, due to the inherent variability and heterogeneity of soil, the range of structure types, and the diversity of loading conditions, SSI remains a challenging topic. The changing climate and escalating environmental concerns add further complexity to SSI and foundation engineering [4–6].

Table 1 offers brief explanations of key terminologies associated with SSI, a critical aspect of this study. **Figure 1** visualizes the SSI concept, illustrating the interaction between structures and soils. This chapter delves into the challenges and complexities of SSI, providing an overview of theoretical aspects, discussing advanced computational models and their applications, and exploring sustainable practices in foundation engineering. Its objective is to contribute to the ongoing dialog in the field, inspire future research, and guide the engineering community through the complexities of SSI.

2. Theoretical aspects of soil-structure interaction

The theoretical aspects of SSI cover a wide range of topics due to the complex interplay between structural and geotechnical engineering principles. Here are some key theoretical aspects:

Term	Definition
Soil-structure interaction (SSI)	The mutual response between a structure and the soil upon which it's built. This dynamic significantly affects the structure's stability and performance [7].
Kinematic interaction	The modification of the soil's natural (free-field) motion due to the presence of a structure. This interaction often affects the seismic response of buildings [8].
Inertial interaction	Inertia of the structure influences the soil's deformation, adding to the complexity of SSI. This interaction can impact the damping characteristics of structures [9].
Free-field motion	The motion that the soil would experience in the absence of a structure. Essential for understanding the unmodified behavior of the soil [10].
Impedance	The complex-valued frequency-dependent force-displacement relation for a foundation. Varies with the type of soil and its characteristics [11].

Table 1.
Key terminologies associated with soil-structure interaction.

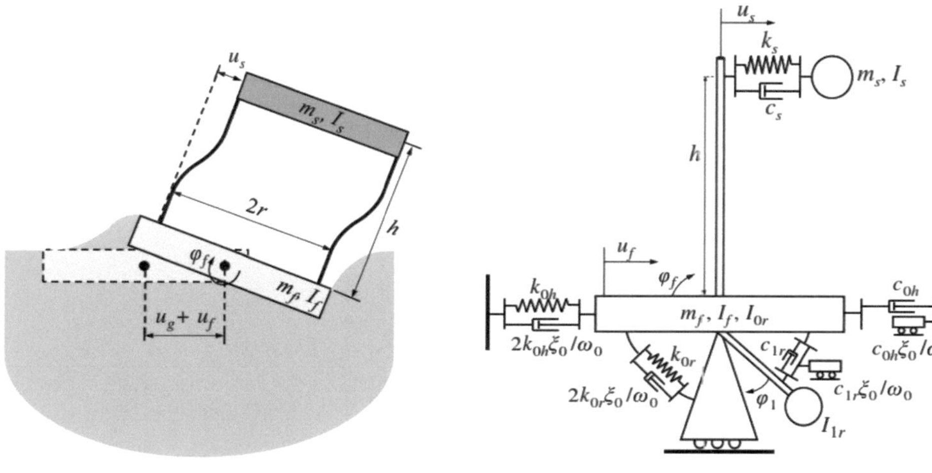

Figure 1.
(a) Soil-structure system; and (b) soil-structure discrete model [12].

i. *Mutual Interaction:* At the core of SSI is the mutual interaction between the structure and the soil. The behavior of the structure impacts the stress–strain state in the soil, while the deformation and movement of the soil influence the structure's behavior.

ii. *Kinematic and Inertial Interaction:* SSI can be broadly categorized into kinematic and inertial interactions. Kinematic interaction pertains to the influence of soil deformation on the structure, while inertial interaction refers to the impact of structural forces on the soil.

iii. *Impact on Structural Response:* SSI can notably affect how structures respond to loads, particularly dynamic ones such as those caused by earthquakes. The effect of SSI becomes particularly pronounced for hefty structures located on relatively soft soils, such as nuclear power plants, high-rise buildings, and elevated highways.

iv. *Nonlinear Behavior:* SSI involves nonlinear behavior in the soil and the contact surface between the structure and the ground. This encompasses nonlinear deformation behavior in the soil and nonlinearity in the interface elements, like sliding and rocking.

v. *Importance in Design:* The effects of SSI can significantly modify the forces and deformations in a structure compared to a fixed-base analysis, which may result in varying design requirements. Consequently, the consideration of SSI is crucial in the design of structures, especially for seismic designs.

The theoretical understanding of SSI is intricate, involving sophisticated numerical modeling techniques and advanced concepts of soil and structural dynamics. Various mathematical models, ranging from simplified linear models to detailed nonlinear ones, are utilized to represent the soil, the structure, and their interaction. These models are typically solved using computational methods like finite element analysis. Several mathematical models are employed to represent SSI, one of which is captured by Eq. (1), illustrating the dynamic equilibrium of a soil-structure system. This equation typically includes the mass, damping, and stiffness matrices of both the structure and soil, in addition to the loading terms. In a simplified matrix form, it can be expressed as:

$$[M]a + [C]v + [K]d = F \quad (1)$$

where:
[M]: is the mass matrix,
a: is the acceleration vector,
[C]: is the damping matrix,
v: is the velocity vector,
[K]: is the stiffness matrix,
d: is the displacement vector,
F: is the force vector.

This equation is typically solved using various numerical methods, such as the Finite Element Method (FEM) or the Finite Difference Method (FDM), to analyze the soil-structure system. The theoretical aspects of SSI encompass a wide array of principles and concepts, developed over many years of research and practical applications. These provide a fundamental understanding of how structures and soil interact and offer guidance for designing and constructing safe and effective structures. SSI theory usually includes the evaluation of stress distribution in soil, the estimation of soil deformation and settlement, and the assessment of soil's dynamic response under varying load conditions. It considers the three-dimensional nature of structures and soil and considers the influence of the soil's nonlinear, anisotropic, and inelastic properties on the behavior of structures [3, 13, 14].

Key considerations in SSI include the type and depth of the foundation, soil stratification and properties, and the type and intensity of loads. For instance, shallow foundations, typically used for light structures, rely on the concept of bearing capacity to ensure structural safety. Conversely, deep foundations like piles and drilled shafts are used for heavier structures or when the top layers of soil are weak. These foundations derive their strength from both side friction and end bearing. Their design and analysis involve understanding the pile-soil interaction under different loading conditions [2, 4, 15]. In the following sections, we will delve deeper into these theoretical aspects, discuss advanced computational models used for SSI analysis, and explore practical considerations and recent research findings in the field. Numerous techniques have been developed for modeling SSI, each with its unique advantages and applications. These methods are summarized in **Table 2**.

3. Soil-structure interaction in seismic design

Soil-structure interaction (SSI) plays a pivotal role in seismic design. During an earthquake, the energy transmitted through the ground interacts with structures, impacting their response. Comprehending and incorporating SSI can notably enhance the accuracy of seismic design and aid in averting structural failures during earthquakes. The performance of structures during seismic events is paramount in civil

Modeling technique	Applications
Finite Element Method (FEM)	Ideal for simulating complex SSI problems with diverse soil properties and geometrical constraints [16].
Boundary Element Method (BEM)	Efficient in modeling unbounded soil regions, reducing the computational effort in large SSI problems [17].
Semi-Analytical Methods	Provides quick solutions for simplified SSI problems, enabling an understanding of the underlying physics without intense computation [18].

Table 2.
Different SSI Modeling techniques and their applications.

engineering. Often, a building's response to such events is dictated not only by the structure itself but by the interaction between the structure and the ground on which it's built - this is known as SSI. SSI can alter the characteristics of the ground motion experienced by a structure during an earthquake, as well as the forces and deformations the structure undergoes [3]. Some essential points about the role of SSI in seismic design include:

i. *Ground motion characteristics:* A building's response to an earthquake relies not only on the characteristics of the building itself but also on the ground motion characteristics. Soil type and properties can significantly impact ground motion characteristics, such as amplitude, frequency content, and duration. Understanding the interaction between the soil and the structure can assist in predicting the ground motion at the site of the structure during an earthquake.

ii. *Influence on structural response:* SSI can affect the structural response during an earthquake. For instance, a structure on soft soil may experience larger displacements and prolonged vibration periods compared to the same structure on stiff soil. These effects can lead to larger seismic forces and deformations in the structure.

iii. *Kinematic and inertial interaction:* In seismic design, SSI can be broadly divided into kinematic and inertial interactions. Kinematic interaction pertains to the effect of soil displacement on the structure, while inertial interaction refers to the influence of the structure's inertia forces on the soil, potentially causing additional soil deformations. Both these effects need to be considered in seismic design.

iv. *Nonlinear behavior:* Earthquakes often induce large strains in both the soil and the structure, leading to nonlinear behavior. This includes nonlinear deformation in the soil and nonlinearity in the interface elements, like sliding and rocking. SSI analysis for seismic design often needs to account for this nonlinear behavior.

v. *Design considerations:* The effects of SSI can result in different design requirements compared to a fixed-base analysis. For example, the increased displacements due to SSI may necessitate a more ductile design, or the heightened forces may require a stronger design. Consequently, taking SSI into account can lead to safer and more economical designs.

For these reasons, SSI is a crucial consideration in the seismic design of structures. However, SSI analysis can be complex, necessitating sophisticated numerical methods and a comprehensive understanding of soil and structural dynamics. Therefore, in practice, SSI is often considered in the design of critical or large structures, like nuclear power plants or major bridges, where the effects of SSI can be significant. For smaller or less critical structures, the effects of SSI are often approximated or neglected, depending on the specifics of the situation.

3.1 Introduction to seismic design

Seismic design refers to the practice of devising structures to ensure their sufficient resistance to seismic activity. The primary objective is safeguarding human life. Traditionally, seismic design principles have been grounded in the concept of strength. This design approach sought to guarantee that the structure could withstand forces induced by ground shaking without collapsing. However, this approach has gradually evolved to accommodate other factors such as structural performance, economic considerations, and the necessity for structures to maintain functionality post-earthquake.

In the process of seismic design, a structure's likely behavior during an earthquake is modeled and analyzed to ascertain its capacity to resist anticipated forces and movements. This process frequently entails considering the soil's behavior under seismic loading conditions and the interaction between the structure and the ground during such events [3, 19]. In this context, SSI assumes a pivotal role. It's the mutual influence that a structure and the supporting soil have on each other during seismic activity. This interaction can significantly impact the structure's response to an earthquake, affecting its safety and performance. Therefore, considering SSI in seismic design is crucial for enhancing the robustness and resilience of the structure against earthquake-induced forces.

3.2 Importance of soil-structure interaction in seismic design

Incorporating soil-structure interaction (SSI) into a seismic design is essential for several reasons [3, 14, 20]:

i. *Site-specific ground motion:* The ground motion at a site during an earthquake is greatly influenced by the local soil conditions. The soil can amplify or dampen seismic waves, impacting the extent of shaking a structure undergoes. This phenomenon, known as site-specific ground motion, can substantially impact a structure's response and needs to be accurately accounted for in a seismic design.

ii. *Alteration of dynamic characteristics:* SSI can change a structure's dynamic properties. The interaction between the structure and the ground can modify the structure's natural frequencies and modes of vibration, which, in turn, influence its response to seismic activity. Neglecting SSI in seismic design can lead to underestimation or overestimation of the structure's response, potentially resulting in unsafe or excessively conservative designs.

iii. *Additional forces and deformations:* SSI can introduce extra forces and deformations in the structure. During an earthquake, the relative movement

Aspect	Neglecting SSI	Considering SSI
Structural response	The structural response may be overestimated, leading to an inaccurate understanding of the structure's behavior under load [8].	SSI consideration leads to a more realistic estimation of the structure's response under different load conditions [9].
Safety margin	Ignoring SSI may overestimate the safety margins, potentially leading to unsafe designs [10].	Including SSI can provide a realistic safety margin, thereby enhancing the reliability of the design [11].
Costs	Overdesign and unnecessary expenditure might result from neglecting SSI [21].	Considering SSI can lead to a more efficient design, potentially reducing construction and maintenance costs [7].

Table 3.
Comparison of the effects of neglecting SSI in design vs. considering it.

between the structure and the ground can induce additional forces in the structure, which need to be considered in the design. Furthermore, ground deformation due to seismic activity can also lead to extra structural deformation.

A comparison of the effects of neglecting SSI in design versus considering it is presented in **Table 3**. This comparison highlights the significance of including SSI in seismic design and the potential risks associated with its omission.

3.3 Challenges in incorporating SSI in seismic design

Incorporating soil-structure interaction (SSI) into a seismic design can be challenging due to its inherent complexity and the resources required for proper analysis. Here are some of the primary challenges [3, 20]:

i. *Complexity of variables and uncertainties:* The behavior of soil during seismic events is influenced by numerous variables such as its stiffness, damping, density, and many others. These properties can significantly vary spatially and with depth. Moreover, the characteristics of seismic ground motion, such as its frequency content, amplitude, and duration, can also widely vary. The abundance of these variables and uncertainties makes it difficult to accurately model and analyze SSI.

ii. *Computational cost:* Given the complexity of SSI, sophisticated numerical methods, like finite element analysis, are often needed for accurate modeling and analysis. However, these methods can be computationally intensive and time-consuming. This is especially the case for large and complex structures or sites with complex soil conditions, which may require extensive computational resources and can prolong the analysis process.

iii. *Lack of comprehensive guidelines:* Although there are various standards and guidelines available, they often provide generic recommendations and lack detailed instructions on how to model and analyze SSI for specific situations. This lack of comprehensive, easy-to-use guidelines can make it difficult for practitioners to incorporate SSI into their designs. Hence, there's a need for more explicit guidelines and standards to aid in the practical application of SSI principles in seismic design.

3.4 Current approaches to address SSI in seismic design

Despite these challenges, there are ongoing efforts to address SSI in seismic design. One prevalent approach is conducting a site-specific ground motion analysis. This process involves modeling the local soil conditions and simulating the ground motion at the site, owing to a range of potential earthquakes. The analysis results are then utilized to derive the seismic input for the structural design. Another strategy is to carry out a dynamic soil-structure interaction analysis. This approach requires modeling both the structure and the soil and simulating their interaction under seismic loading. This process enables the calculation of additional forces and deformations in the structure due to SSI, which can be incorporated into the design. Several methodologies are also under development to simplify the modeling and analysis of SSI. For instance, substructure methods allow for independent analysis of the structure and the soil, thereby reducing computational costs. In addition, surrogate modeling techniques are being developed to approximate the complex behavior of SSI, thereby rendering its analysis more manageable [3, 14].

Table 4 presents an overview of some of the current strategies to address Soil-Structure Interaction (SSI) in seismic design. It is crucial to note that each of these strategies has its own advantages and limitations. The choice of method will depend on the specifics of the structure and site, the structure's significance, and the resources available for the analysis. In practice, more than one method might be employed. For instance, a simplified method might be used for preliminary design, while a more detailed method could be used for the final design. Moreover, these methods are continuously being developed and improved as our understanding of SSI and computational capabilities grow.

3.5 Future directions in SSI for seismic design

While significant progress has been made in addressing SSI in seismic design, there remains a substantial amount of work yet to be accomplished. Future research should focus on enhancing the precision and efficiency of SSI modeling and analysis and developing comprehensive, practical guidelines for its inclusion in seismic design. Additionally, there is a growing need for more in-depth research into the effects of SSI on non-structural elements of buildings, such as partitions, ceilings, and mechanical and electrical systems. These elements can be significantly impacted by SSI, and their failure could pose a risk to human safety and disrupt the building's functionality. Lastly, it's essential to better incorporate SSI considerations into the broader framework of performance-based seismic design. This approach involves designing structures to meet specific performance objectives, such as life safety, building functionality, and economic loss, under various levels of seismic hazards. The inclusion of SSI in this framework can lead to more realistic and effective seismic designs.

Approach	Description	Advantages	Limitations
Direct method	It involves modeling the entire soil-structure system and applying the seismic input at the base of the soil layer. It is a comprehensive approach where the structure and the soil are modeled together in a single analysis.	Most accurate representation of SSI. Can account for complex soil-structure configurations and soil behaviors.	Requires significant computational resources. Requires detailed soil and structural information.
Substructure method	Divides the problem into two separate analyses: the free-field soil response and the structure-soil interaction. The results from the free-field response analysis are used as input for the structure-soil interaction analysis.	Less computationally intensive than the direct method. Allows for separate consideration of soil and structure, which can be advantageous in some cases.	Less accurate representation of SSI. Cannot account for some complex soil-structure configurations.
Equivalent linearization method	Assumes linear elastic behavior of the soil and the structure but incorporates the effects of soil damping. It is an iterative process where the soil properties are updated based on the strain level.	Less computationally intensive than fully nonlinear methods. Provides a good approximation for small to moderate strain levels.	Not accurate for large strain levels where nonlinear soil behavior becomes significant.
Simplified methods	These methods involve simplifications such as considering the structure as a rigid block or considering the soil as a set of springs and dashpots. These methods are often used for preliminary design or for structures where the effects of SSI are not critical.	Requires less computational resources. Easier to implement.	Less accurate representation of SSI. Cannot account for complex soil-structure configurations and soil behaviors.

Table 4.
Overview of current approaches for addressing soil-structure interaction (SSI) in seismic design.

4. Evaluating soil-structure interaction

Having established a fundamental understanding and appreciation of SSI, the subsequent phase concentrates on evaluation methods. The evaluation of SSI incorporates both experimental and computational approaches. These techniques assist in gaining insights into the intricate behavior of the soil and the structure during seismic events, laying the foundation for the design and assessment of structures subjected to such events [22]. Furthermore, evaluating SSI involves discerning how a structure interacts with the soil on which it stands, particularly during occurrences such as earthquakes. This evaluation is critical for seismic design as it can substantially influence a structure's behavior and stability during seismic events. The key steps involved in evaluating SSI are as follows:

- *Soil and site characterization*: The initial step in evaluating SSI is understanding the soil and site characteristics. This involves geotechnical investigations to determine the soil type, layering, and properties like stiffness, density, and damping. For seismic design, additional details such as the soil's shear wave velocity may be required.

- *Structural characterization*: It's necessary to define the properties of the structure, like its mass, stiffness, and damping. For seismic design, the dynamic characteristics of the structure, such as its natural frequencies and mode shapes, are particularly important.

- *Interaction analysis*: Subsequently, analyzing the interaction between the structure and the soil becomes the next crucial step. This may involve various methods, ranging from simplified to more sophisticated ones that model the entire soil-structure system. The choice of method will rely on the specifics of the structure and site, the structure's importance, and the resources available for analysis.

- *Interpretation of results*: The interaction analysis results need interpretation to understand their implications for the structure's design. This could involve checking the structural response against relevant design criteria, considering the effects of the soil-structure interaction on the structural response, and assessing the need for any mitigation measures.

- *Design iteration*: Depending on the interaction analysis results, it may be necessary to iterate the design of the structure. This could involve changes to the structural design, the foundation design, or even the site preparation.

- *Verification and validation*: The final step is to verify and validate the interaction analysis. This could involve comparing the predicted behavior with observed behavior for similar structures or sites or using other forms of verification such as peer review or additional analyses.

By meticulously evaluating SSI, engineers can design structures that are more resilient to seismic events, leading to safer and more sustainable built environments.

4.1 Experimental approaches

Experimental approaches to SSI evaluation largely include laboratory testing and field testing. Laboratory testing involves small-scale models of soil and structures which are subjected to simulated seismic loading conditions. This method provides invaluable information on the behavior of the soil, its structure, and its interaction under controlled conditions. On the other hand, field testing involves full-scale structures and actual soil conditions. Techniques such as seismic shaking table tests and vibration tests are commonly employed. While field testing provides more realistic data, it's important to note that the complexity and costs associated with these tests can be substantial [23, 24]. The choice of experimental approach hinges on factors such as the specifics of the structure and site, the research question or design issue under consideration, and the resources available for testing. In practice, both experimental and analytical methods are often utilized in tandem for a comprehensive evaluation of SSI (**Table 5**).

Experimental approach	Description	Advantages	Limitations
Shake table tests	These tests involve constructing a scale model of the structure and the soil and subjecting it to shaking that simulates earthquake ground motions. The responses of the structure and the soil are then measured.	Can simulate realistic earthquake ground motions. Can observe the actual physical behavior of the soil-structure system.	Scale effects can influence the results. May be difficult or costly to reproduce some soil conditions.
Centrifuge tests	These tests use a centrifuge to apply a high gravitational field to a scale model of the soil-structure system. This simulates the stress conditions in the actual soil-structure system.	Can reproduce the actual stress conditions in the soil-structure system. Can observe the actual physical behavior of the soil-structure system.	Scale effects can influence the results. May be difficult or costly to reproduce some soil conditions and structural responses.
Field tests	These tests involve actual structures and soils. They may involve inducing vibrations in the structure and measuring the response or measuring the respo3nse of the structure and the soil during actual earthquakes.	Can provide data on the behavior of actual soil-structure systems. Can capture the effects of actual soil conditions and structural responses.	Can be expensive and logistically challenging. May be difficult to control or predict the loading conditions.
Laboratory soil tests	These tests involve testing soil samples in the laboratory to measure properties such as stiffness, damping, and shear strength. The results can be used in the analysis of the soil-structure interaction.	Can provide detailed information on soil properties. Can control the testing conditions.	May not fully represent the actual soil conditions in the field. May be difficult to capture the interaction effects with the structure.

Table 5.
Experimental approaches for evaluating soil-structure interaction: Overview, advantages, and limitations.

4.2 Computational approaches

Computational approaches to SSI evaluation involve utilizing numerical methods to simulate the soil and the structure's behavior, as well as their interaction under seismic loading. The Finite Element Method (FEM) and the Finite Difference Method (FDM) are widely employed due to their versatility in modeling intricate geometries and materials. However, these methods can be computationally demanding, particularly for large-scale or complex problems. As a result, simplified methods such as the substructure method and the equivalent linearization method are often employed for practical design applications. These methods can approximate the behavior of SSI with an acceptable degree of accuracy, while significantly reducing computational costs [22, 25]. The choice of computational approach is dependent on the specifics of the problem, which includes the nature of the soil and the structure, the type of loading, and the level of detail required in the analysis (**Table 6**).

Computational approach	Description	Advantages	Limitations
Finite element method (FEM)	A numerical technique for finding approximate solutions to boundary value problems for differential equations. It subdivides a large system into smaller, simpler parts that are called finite elements.	Capable of modeling complex geometries and heterogeneous materials. Can handle nonlinear behaviors.	Computationally intensive, especially for large systems or for nonlinear problems. Requires a careful selection of element types and mesh refinement.
Boundary element method (BEM)	A numerical computational method of solving linear partial differential equations which have been formulated as integral equations. It can be applied in many areas of engineering and science including fluid mechanics, acoustics, electromagnetics, and fracture analysis.	Less computationally intensive than FEM for many problems. Only the boundary of the region needs to be meshed.	Not well-suited for problems involving non-linearities or inhomogeneities.
Finite difference method (FDM)	A numerical method for solving differential equations by approximating derivatives with finite differences.	Simple to implement and understand. Less computationally intensive than FEM for many problems.	May not handle complex geometries or heterogeneous materials as well as FEM.
Discrete element method (DEM)	A numerical technique that calculates the interaction of a large number of particles (like soils).	Good for problems with large displacements and rotations, and for problems where the structure of the material evolves during loading (like granular soil).	Can be computationally intensive for systems with a large number of particles.

Table 6.
Computational approaches for evaluating soil-structure interaction: Overview, advantages, and limitations.

4.3 Challenges in evaluating SSI

Despite the availability of these approaches, the evaluation of SSI is laden with difficulties. The complexity of SSI, stemming from the interaction between the nonlinear, hysteretic behavior of the soil, and the dynamic behavior of the structure, makes it a daunting task to model and analyze accurately. Furthermore, uncertainties in soil properties and seismic ground motion can significantly influence the results of SSI evaluation. Consequently, there is a pressing need for methods that can adequately accommodate these uncertainties [26]. However, the evaluation of SSI is a complex endeavor encompassing various challenges, including but not limited to:

1. *Modeling complexity*: SSI involves the interaction between diverse types of materials such as concrete, steel, and various types of soil, which may exhibit non-linear, rate-dependent, and path-dependent behavior. Accurately modeling these materials and their interactions can be quite challenging.

2. *Geometric complexity*: Structures and their foundations can possess complex geometries, and the soil layers themselves can also display geometric complexity, particularly at sites with irregular topography or stratigraphy. This geometric complexity can make the analysis more difficult.

3. *Loading complexity*: Loads such as earthquakes, wind, and traffic can have intricate temporal and spatial distributions. In the case of earthquakes, the motion can occur in all three dimensions (horizontal, vertical, and rotational), which adds to the problem's complexity.

4. *Uncertainty*: Significant uncertainty often exists in the properties of the materials (particularly the soil), the loading conditions, and the modeling parameters. This uncertainty can make it challenging to make accurate predictions of the system's behavior.

5. *Computational demands*: Analyses of SSI can be computationally demanding, especially for large or complex systems or when non-linear material behavior is incorporated into the model. This can limit the feasibility of performing detailed analyses, particularly for routine design work.

6. *Validation of models*: Validating the computational models used for SSI analysis can be challenging due to the difficulties in performing full-scale tests and the uncertainties involved in interpreting the results of such tests.

These challenges imply that the evaluation of SSI often necessitates a blend of sophisticated computational modeling, careful interpretation of field and laboratory test data, and sound engineering judgment. Despite these hurdles, the importance of SSI in many engineering problems underscores its significance as a vital area of study and research.

4.4 Recent advances in SSI evaluation

Recent advancements in computational technology and methods have opened new avenues for the evaluation of SSI. High-performance computing has paved the way for more intricate and realistic simulations of SSI. Concurrently, progress in numerical methods, such as the development of non-linear soil models and stochastic analysis methods, have heightened the accuracy and dependability of SSI evaluation. Additionally, innovative experimental techniques like micro-electro-mechanical systems (MEMS) sensors and digital image correlation (DIC) have refined the precision and expanded the reach of SSI measurements, thereby augmenting the data available for validating computational models [26]. While these advancements offer promising avenues, they simultaneously underscore the importance of sustained research and development to further refine the state-of-the-art in SSI evaluation.

5. Mitigating challenges in soil-structure interaction

Understanding and evaluating SSI is more than just an academic endeavor; its fundamental aim is to inform the design and construction of structures capable of efficiently withstanding seismic events. Consequently, mitigating the challenges associated with SSI carries paramount significance. These mitigation strategies can encompass everything from sophisticated modeling techniques to innovative construction methodologies.

5.1 Refined modeling techniques

Indeed, accurately encapsulating the intricate behavior of SSI poses significant challenges. Therefore, there is an ongoing effort to refine existing models and to devise new ones. Recent advancements in numerical modeling techniques, such as nonlinear dynamic analyses, probabilistic methods, and hybrid simulations, have shown considerable promise in delivering more accurate predictions of SSI. For example, hybrid simulation techniques—where a portion of the system is modeled numerically and the rest experimentally—have emerged as a popular method to analyze complex SSI problems. This approach strikes a balance between the realism offered by experimental tests and the adaptability inherent in numerical simulations [27, 28].

5.2 Innovative construction methods

Apart from advancements in modeling techniques, innovative construction methods and technologies have also been developed to mitigate the effects of SSI. Techniques such as ground improvement, isolation systems, and energy dissipation devices are now commonly used to enhance the performance of structures subjected to seismic loading. Ground improvement techniques aim to enhance the properties of the soil, reducing its potential to amplify seismic motions. These methods include compaction, grouting, and soil stabilization, among others. Conversely, isolation systems and energy dissipation devices are installed within the structure to minimize the forces transmitted from the ground to the structure during an earthquake. Examples of these systems include base isolation mechanisms and dampers [29, 30].

5.3 Design codes and guidelines

Efforts to mitigate the challenges of SSI are also reflected in various design codes and guidelines. These resources provide practitioners with practical methods and criteria for incorporating SSI considerations into the design and assessment of structures. However, these codes and guidelines are based on a simplified understanding of SSI and may not fully account for its complexity. Therefore, ongoing research and development are essential to enhance these standards and ensure they reflect the most recent understanding and advancements in the field of SSI [26]. **Table 7** provides an outline of some common design codes and guidelines that address soil-structure interaction.

5.4 Soil improvement and reinforcement techniques

Soil improvement and reinforcement techniques play a pivotal role in managing the risks associated with SSI. These techniques aim to enhance the soil's inherent

Code/guideline	Originating organization	Description	Application
Eurocode 8: Design of structures for earthquake resistance	European Committee for Standardization (CEN)	Part of the Eurocode series, Eurocode 8 deals with the design and analysis of structures subject to seismic actions, including considerations of SSI.	Used in European Union countries for the seismic design of structures, including the consideration of SSI.
ASCE 7: Minimum Design Loads for Buildings and Other Structures	American Society of Civil Engineers (ASCE)	This standard provides requirements for general structural design and includes means for determining dead, live, soil, flood, wind, snow, rain, and earthquake loads, as well as their combinations. It includes provisions for considering SSI effects.	Widely used in the United States for the design of buildings and other structures.
NEHRP Recommended Seismic Provisions	National Earthquake Hazards Reduction Program (NEHRP), FEMA	These provisions include criteria for the seismic design of structures. It includes considerations for SSI and is intended to be adopted by model building codes.	Used as a reference in the United States for seismic design, including SSI considerations.
ACI 318: Building Code Requirements for Structural Concrete	American Concrete Institute (ACI)	This code provides minimum requirements for the materials, design, and detailing of structural concrete buildings and, where applicable, nonbuilding structures, including considerations for SSI in the foundations.	Widely used internationally for the design of concrete structures.
Canadian Standard CAN/CSA S6-14: Canadian Highway Bridge Design Code	Canadian Standards Association (CSA)	This standard provides requirements for the design, evaluation, and structural rehabilitation of fixed and movable highway bridges in Canada, including considerations for SSI.	Used in Canada for the design of highway bridges.
Japanese Building Standard Law - Seismic Design Standards	Building Center of Japan	Covers the requirements for seismic design in Japan, including considerations for soil-structure interaction.	Used in Japan for the seismic design of buildings and infrastructure.
IS 1893 (Part 1): Criteria for Earthquake Resistant Design of Structures	Bureau of Indian Standards (BIS)	Part of Indian Standards, this code deals with the assessment of seismic loads on structures and covers general provisions and buildings, including considerations for SSI.	Used in India for the seismic design of structures.

Table 7.
Key design codes and guidelines addressing soil-structure interaction (SSI).

properties, thereby making it more resistant to deformation and failure under various loads, including those induced by seismic activities [31, 32]. Some commonly used techniques include:

i. *Deep soil mixing (DSM)*: This method involves the mechanical blending of soil with cementitious materials using a specially designed auger. The improved soil boasts enhanced strength and reduced permeability, making it more resistant to seismic shaking and other loads.

ii. *Vibro-compaction*: This process densifies loose, granular soils. It involves a vibrating probe inserted into the soil. The vibration causes the soil particles to rearrange and become denser, thus increasing the soil's shear strength, and reducing the risk of liquefaction during a seismic event.

iii. *Dynamic compaction*: This technique compacts the subsoil strata using a high-energy impact from a dropped weight. This process enhances soil properties by reducing voids and increasing density, thereby improving the soil's resistance to seismic waves.

iv. *Geosynthetic reinforcements*: These include geotextiles, geogrids, and geocells used to reinforce the soil. Typically used in the construction of retaining walls, slopes, embankments, and road construction, these materials distribute loads over a wider area, thus increasing the soil's shear strength and stability.

These soil improvement and reinforcement techniques can significantly reduce the effects of SSI by bolstering the soil's seismic resistance. However, the choice of the appropriate technique depends on several factors, including the type of soil, the load to be supported, and the specific requirements of the project.

5.5 Advanced foundation design

The design of a foundation plays a crucial role in mitigating the effects of SSI. Advanced foundation design approaches such as pile foundations, mat foundations, and raft foundations are frequently employed. These types of foundations distribute loads over a larger area, thereby reducing the stresses transmitted to the soil and minimizing the potential for excessive deformation and failure (Das 2016). A summary of these foundation types is as follows:

i. *Pile foundations*: Pile foundations are long, slender, columnar structures driven into the ground to support loads. They are particularly useful in transferring loads from a structure through weak, compressible strata or water onto stronger, more compact, less compressible, and stiffer soil or rock at depth. They also aid in resisting uplift forces due to seismic activity and other factors.

ii. *Mat foundations*: Also known as raft foundations, these are large, continuous concrete slabs that support the entire weight of the structure. They distribute the load over the entire footprint of the building, reducing the stress on the soil. Mat foundations are beneficial when dealing with weak soils, as they evenly distribute the load across a larger surface area, thus reducing the potential for excessive deformation and failure.

iii. *Raft foundations*: A raft foundation is a thick concrete slab reinforced with steel that covers the entire contact area of the structure, much like a thick floor. Sometimes the area covered by the raft may be greater than the contact area, depending on the soil's bearing capacity underneath. The reinforcing bars usually run perpendicular to each other in both the top and bottom layers of steel reinforcement.

These advanced foundation designs, when appropriately used, can offer a more effective and resilient response to the challenges posed by SSI. The choice among these foundation types of hinges on various factors, including the nature of the load, the type of structure, the properties of the soil, and the presence of any potential seismic activity.

5.6 Earthquake resistant design

Earthquake-resistant design principles aim to ensure that structures can withstand the forces generated by earthquakes without sustaining significant damage. This involves designing the structure to have enough strength, stiffness, and ductility to resist seismic forces. The key principle of earthquake-resistant design is to ensure that the structure can deform in a controlled manner under seismic loading, thereby preventing catastrophic failure. Furthermore, seismic base isolation and energy dissipation techniques are widely used in earthquake-prone areas to decrease the forces transmitted from the ground to the structure. These techniques aid in mitigating the impacts of SSI in the following ways:

i. *Seismic base isolation*: This method decouples the structure from the ground, thereby reducing the amount of energy that the structure absorbs during an earthquake. Isolators are often installed between the structure and the ground and are designed to deform and absorb seismic energy, limiting the forces transferred to the structure. Common types of isolators include elastomeric bearings, sliding bearings, and friction pendulum bearings.

ii. *Energy dissipation techniques*: These techniques involve the use of devices that absorb and dissipate the energy induced by seismic events, reducing the amount of energy transferred to the structure. Common energy dissipation devices include viscous dampers, friction dampers, and metallic yield dampers.

By isolating the structure from the ground motion or dissipating the seismic waves' energy, these techniques help to reduce the impact of SSI. However, it is important to note that these techniques should be applied in conjunction with other design practices to effectively mitigate the potential impacts of earthquakes [29, 33].

5.7 Soil-structure interaction: mitigation and management

Managing the effects of SSI necessitates the implementation of a comprehensive strategy that considers all the aforementioned factors. This strategy may include thorough site investigation, sophisticated geotechnical analysis, appropriate structural design, and the employment of advanced technologies for soil improvement and reinforcement. The role of numerical modeling is also vital in managing SSI. Advanced numerical models that can simulate the behavior of both soil and structure under different loading conditions can provide valuable insights into the potential impacts of SSI. These insights can guide the design and construction process, helping engineers to mitigate the effects of SSI [34]. In conclusion, while the challenges posed by SSI are considerable, they can be effectively managed through a combination of advanced engineering techniques and innovative design practices. Continuous research and development in this area will enable engineers to further their understanding of SSI and develop more effective solutions to mitigate its effects. The field of SSI is a vital area of study that holds significant potential for improving the resilience and safety of structures, especially in regions prone to seismic activity.

6. Case studies

In this section, we present a series of case studies that underscore the challenges posed by SSI in real-world projects, as well as the approaches used to tackle them. These case studies provide concrete examples of the principles and techniques previously discussed. An array of case studies has been conducted to understand the implications of SSI in real-world circumstances. The primary findings from these case studies are consolidated and presented in **Table 8**.

6.1 Case study 1: sustainable practices in soil-structure interaction

In the recent years, the push towards sustainable construction practices has extended into the realm of geotechnical engineering, encompassing soil-structure interaction (SSI) too. This has resulted in the creation and application of eco-friendly materials and groundbreaking techniques that not only secure structural safety and longevity, but also minimize the environmental impact of construction activities. A notable illustration of such sustainable practices is the utilization of recycled materials in geotechnical applications. The use of recycled construction and demolition waste (CDW) as backfill, or reinforcement materials can considerably diminish the demand for traditional, non-renewable geotechnical materials like sand or gravel. In addition,

Case study	Key findings
SSI in Taipei 101, Taiwan [35]	Showcased the impact of SSI on the high-rise building's dynamic behavior and highlighted the role of deep soft soil layers in amplifying ground motions.
SSI effects in the design of Zakim Bridge, Boston [36]	Identified the importance of SSI, particularly the role of pile group effects, in bridge design.
SSI in the seismic analysis of the Fukushima Daiichi Nuclear Power Plant [37]	Underlined the criticality of considering SSI for accurate seismic safety assessments of nuclear power plants.

Table 8.
Major case studies involving SSI and key findings.

studies have shown that the mechanical properties of properly processed CDW can rival, or even exceed, those of traditional geotechnical materials [38, 39].

6.2 Case study 2: computational models in soil-structure interaction

The realm of soil-structure interaction has been revolutionized by advancements in computational power and technology, ushering in more sophisticated and accurate computational models to forecast the behavior of soils and structures under varying conditions. These models facilitate the comprehension of complex SSI phenomena and offer a quantitative foundation for designing and assessing geotechnical engineering solutions. Finite element methods (FEM) and boundary element methods (BEM) are frequently employed for analyzing SSI problems. These numerical methods allow engineers to replicate the intricate behavior of soils and structures within a computer environment, yielding insights into their interaction under diverse load conditions. Other novel computational models, such as the distinct element method (DEM), are also gaining popularity due to their ability to simulate the granular nature of the soil more accurately. These advancements in computational modeling are crucial in designing safer, more efficient structures while minimizing the costs and uncertainties associated with traditional, empirical design approaches [25, 40].

6.3 Case study 3: a highway embankment

The third case study pertains to the construction of a highway embankment over soft, compressible soil as part of an expansive highway improvement project. Given the potential for considerable settlement and instability under the embankment's weight, this posed significant challenges. The project team utilized an innovative approach, which incorporated a blend of ground improvement techniques to augment the soil's properties. The team employed preloading to hasten the consolidation of the soft soil and used geosynthetics to reinforce the soil and enhance its shear strength [41, 42].

The project also gained from a comprehensive geotechnical investigation, which allowed the team to fine-tune the design of the ground improvement works and ensure their effectiveness. Numerical modeling aided in understanding the soil's behavior under the embankment load. Throughout the project, the team vigilantly monitored the settlement and stability of the embankment. This proactive approach enabled the identification of potential issues early on and the implementation of necessary corrective measures [31, 43, 44]. The project reached a successful completion, exhibiting the embankment's performance under traffic loads. This case study emphasizes the importance of understanding and managing SSI during the design and construction of transportation infrastructure.

7. Future directions and concluding remarks

Understanding the intricacies of soil-structure interaction (SSI) is integral to the design and construction of secure and reliable civil engineering structures. As showcased in the case studies in this chapter, SSI management requires a multidisciplinary approach. This involves marrying geotechnical investigations, advanced numerical modeling, inventive foundation solutions, and continuous performance monitoring. Looking into the future, the challenges associated with SSI will undoubtedly continue to develop. Climate change, urbanization, and technological advancements are just some of the factors that will drive these changes. Engineers will need to stay abreast of these developments and adjust their practices accordingly [31].

There is a vast potential for leveraging advanced technologies to improve our comprehension and management of SSI. These technologies include sensors for real-time monitoring of soil and structural behavior, advanced geotechnical investigation techniques, and the application of artificial intelligence and machine learning to optimize foundation design and performance prediction. Sustainability is another key trend in geotechnical engineering. It will be incumbent on engineers to consider the environmental and social impacts of their projects and seek ways to minimize these through sustainable design and construction practices. In conclusion, the study and management of SSI represent a dynamic and challenging field in civil engineering. However, by embracing the complexities of SSI, engineers can devise innovative and sustainable solutions that are able to respond to the demands of our rapidly evolving world [6, 45].

Green technologies and practices are gaining traction in the construction industry, and it is expected that this trend will foster the adoption of eco-friendly foundation solutions. For example, the use of recycled materials and geosynthetics can enhance soil properties and decrease the demand for concrete and steel, materials associated with high embodied energy. The incorporation of Building Information Modeling (BIM) and 3D printing in construction could also transform the way we design and build foundations. These technologies allow for increased precision and efficiency in the design process and could potentially facilitate the construction of more complex and optimized foundation systems [46].

Furthermore, ongoing research and development are necessary in the domain of numerical modeling of SSI. Although current models have significantly advanced our understanding of SSI, they still have limitations in accurately representing complex soil behaviors and interactions under varying loading and environmental conditions. Ultimately, deepening our understanding of SSI and continuously improving the techniques and technologies for managing it are crucial to surmounting the geotechnical challenges of future infrastructure projects. Through continued research and the exchange of knowledge and best practices, we can anticipate a future where the complexities of SSI are embraced and effectively managed, resulting in safer and more sustainable built environments [31, 41, 45, 47].

8. Case studies: challenges and sustainable practices

Case studies offer an essential look into the practicalities of managing soil-structure interaction (SSI). In this section, we will delve into a series of case studies. Each one will spotlight a distinctive challenge related to SSI, alongside the sustainable practices that were effectively implemented to surmount these obstacles.

8.1 Case study 1: managing differential settlement in skyscraper construction

Building skyscrapers introduces numerous challenges for foundation engineering, including managing differential settlements. This was a significant issue confronted during the construction of the Burj Khalifa in Dubai, the tallest building globally at its completion. Towering at an awe-inspiring height of 828 meters, the Burj Khalifa's enormous scale necessitated a solid foundation design. The building's foundation system employed a piled raft, a decision based on comprehensive geotechnical investigations that included both in-situ and laboratory tests. The soil-structure interaction was complex due to the varying soil conditions, comprising layers of sand and weathered rock. To address this complexity, advanced numerical models were deployed to simulate the SSI and predict the settlement behavior of the super-tall structure accurately [48]. These models considered various factors, such as Dubai's calcareous soil's non-linear behavior and the immense vertical loads resulting from the building's weight. The successful completion of the Burj Khalifa represented a significant achievement in foundation engineering, demonstrating modern engineering's capacity to manage complex soil-structure interactions.

8.2 Case study 2: overcoming liquefaction risk in seismically active regions

Soil liquefaction poses serious threats to structures in seismically active regions. This was clearly evident during the 2011 Christchurch earthquake in New Zealand, which led to substantial structural damage due to soil liquefaction. In the aftermath of this catastrophe, comprehensive research was undertaken to improve our understanding and mitigation of the effects of soil liquefaction. This effort involved meticulous investigations into the SSI during and after the seismic event. The ground improvement techniques employed, such as stone columns and deep soil mixing, demonstrated success in reducing the risk of liquefaction and restoring the strength and stiffness of the liquefied soil. Consequently, these techniques have been recommended for deployment in similar seismic scenarios in the future [49]. This case study exemplifies how a combination of geotechnical engineering practices, underpinned by a thorough understanding of soil-structure interaction, can protect structures and, most importantly, save lives in regions prone to seismic activity.

8.3 Case study 3: implementing green foundations in urban construction

Soil liquefaction poses serious threats to structures in seismically active regions. This was clearly evident during the 2011 Christchurch earthquake in New Zealand, which led to substantial structural damage due to soil liquefaction. In the aftermath of this catastrophe, comprehensive research was undertaken to improve our understanding and mitigation of the effects of soil liquefaction. This effort involved meticulous investigations into the SSI during and after the seismic event. The ground improvement techniques employed, such as stone columns and deep soil mixing, demonstrated success in reducing the risk of liquefaction and restoring the strength and stiffness of the liquefied soil. Consequently, these techniques have been recommended for deployment in similar seismic scenarios in the future [49]. This case study exemplifies how a combination of geotechnical engineering practices, underpinned by a thorough understanding of soil-structure interaction, can protect structures and, most importantly, save lives in regions prone to seismic activity.

8.4 Case study 4: dealing with expansive soils

Dealing with expansive soils, which experience considerable volume changes due to variations in moisture content, is a frequent challenge in geotechnical engineering. These soils can inflict significant damage on foundations and structures. A case in point was encountered during the construction of a residential development in Dallas, Texas, where expansive clay soil presented substantial risks. The strategy for mitigation involved designing a stiffened slab foundation system capable of tolerating the volumetric changes of the underlying expansive soil. Furthermore, site-specific moisture conditioning of the soil was undertaken to curtail future volume changes. During the design and execution stages of the foundation system, careful considerations of SSI were paramount. The successful completion of this project underscored the importance of understanding and accommodating SSI when contending with expansive soils [50].

8.5 Case study 5: bridge foundations in flood-prone areas

Constructing bridges in flood-prone areas demands special attention due to the potential impact of scour on the stability of foundations. The challenges related to soil-structure interaction (SSI) were starkly exhibited during the design of the foundation for a bridge across the River Elbe in Germany. To ensure stability against lateral loads due to potential scour effects, the bridge's foundation was designed using large-diameter bored piles. The design process necessitated complex numerical modeling to understand SSI under the combined influence of vehicular loads, river flow, and potential scour. The successful completion of this project highlighted the pivotal role of SSI in the design and construction of bridge foundations in demanding environments [51]. These case studies underscore that understanding SSI is crucial in managing diverse challenges in foundation engineering, which can differ significantly based on the specific attributes of the site and the structure.

9. Sustainable practices in foundation engineering

In response to growing environmental concerns and the urgent call for sustainable development, the sphere of foundation engineering has begun to embrace a multitude of green practices. This section offers an overview of such sustainable practices within foundation engineering and explores their potential to alleviate diverse challenges associated with soil-structure interaction (SSI).

9.1 Use of recycled and natural materials

The use of recycled materials and natural fibers in foundation construction presents an effective approach to decreasing environmental impact. These materials, being widely available and cost-effective, lessen our dependence on non-renewable resources. For example, recycled concrete aggregates (RCA) have been successfully utilized as substitutes for natural aggregates in constructing various types of foundations. Additionally, natural fibers, such as coir, jute, and bamboo, have shown promising results in enhancing soil stability, which in turn improves the performance of foundations constructed on these stabilized soils [52, 53].

9.2 Energy-efficient foundation systems

Energy piles, an innovative concept that merges the structural role of foundation piles with the role of heat exchangers for ground-source heat pumps, exemplify sustainable and energy-efficient foundation systems. By incorporating both the functions of heat exchange and load bearing, energy piles offer a sustainable solution for space heating and cooling, while concurrently ensuring the structural safety of buildings [54].

9.3 Low-impact construction techniques

Low-impact construction methods, like the use of helical piles and micro piles, can significantly minimize environmental disruption during foundation construction. These methods require fewer materials and cause minimal soil disturbance, making them a more sustainable choice for foundation construction, especially in environmentally sensitive areas [55].

9.4 Biotechnical solutions

Biotechnical solutions, another aspect of sustainable foundation engineering, focus on harnessing natural processes for soil stabilization and enhancement. The most recognized biotechnical solution is bio-cementation, a process that employs microorganisms to instigate calcite precipitation within soil, thereby bolstering its strength and reducing permeability [56]. This approach shows promise not only from a sustainability standpoint but also for its potential to augment the performance of soil-structure systems.

9.5 Reuse of excavated soil

Another sustainable practice in foundation engineering is the repurposing of excavated soil. Instead of treating this soil as waste, it can be processed and reused in construction projects. A range of soil improvement techniques, such as soil mixing and stabilization with cementitious materials, can enhance the properties of the excavated soil and make it suitable for reuse [57]. In conclusion, sustainable practices in foundation engineering have immense potential for addressing various challenges related to soil-structure interaction. As advancements in this field continue to unfold, the integration of sustainable principles into foundation engineering is expected to become increasingly prevalent in the future.

10. Future research directions and conclusion

As we progress in our endeavor for sustainable solutions within the field of soil-structure interaction, numerous promising avenues for future research emerge. These are fueled by the escalating need for sustainable urban development and the mounting challenges induced by climate change and resource scarcity.

i. *Sustainable materials*: There is a crucial need for ongoing research into sustainable and recycled materials suitable for use in foundation construction. Such research could not only broaden our array of sustainable materials but also enhance our understanding of their behavior in diverse geotechnical applications.

ii. *Energy efficiency*: Investigating foundations' potential as a source of renewable energy, such as further developing and optimizing energy piles, is an encouraging research trajectory. These technologies could enhance the energy efficiency of buildings and aid in climate change mitigation.

iii. *Biotechnical engineering*: The further development of biotechnical solutions for soil improvement and stabilization could lead to more sustainable and environmentally friendly foundation engineering practices. This might encompass expanding the range of organisms used in bio-cementation and exploring other biological processes that could contribute to soil improvement.

iv. *Advanced computational modeling*: Continued improvements in computational models for soil-structure interaction can help us design more efficient and resilient foundation systems. As these models become more accurate and versatile, they will enable us to better predict and manage the complex behavior of soil and structures.

The implementation of sustainable practices in soil-structure interaction is an imperative step towards a more resilient and sustainable urban future. As our understanding and technologies advance, we are likely to witness an even greater emphasis on sustainability in the field of foundation engineering.

10.1 Advanced geotechnical investigation techniques

The need for advanced geotechnical investigation techniques to comprehend the phenomena of soil-structure interaction more effectively is paramount. These include the employment of non-destructive testing methods, such as geophysical methods—seismic refraction and electrical resistivity, for instance—that can provide comprehensive data on subsurface conditions without disturbing the soil [58]. Furthermore, remote sensing technologies, which utilize satellite or airborne data to assess and monitor the properties of the soil and the site, offer significant potential for broad, high-resolution site characterization. These sophisticated techniques facilitate a more holistic understanding of the soil's properties and behavior, thus enabling more accurate prediction and management of soil-structure interactions.

10.2 Development of innovative materials

Continued research into the development of innovative materials for soil improvement and foundation construction is a crucial aspect of future exploration. This includes harnessing the potential of recycled and waste materials to reduce environmental impact and reliance on virgin resources. Bio-inspired materials, inspired by nature and its mechanisms, also present a rich source of potential solutions. Simultaneously, the advancement and optimization of geosynthetics—synthetic products used to stabilize terrain—represent another promising area of research. These materials not only offer opportunities for enhanced performance but also contribute to the sustainability of construction practices [59].

10.3 Machine learning and AI in foundation engineering

The integration of artificial intelligence (AI) and machine learning (ML) in foundation engineering heralds a promising future research direction. These advanced digital tools can aid in the accurate prediction of soil behavior, the optimization of foundation design, and ongoing performance monitoring of established structures [60]. In conclusion, the realm of soil-structure interaction is in a state of constant evolution, stimulated by the mounting demand for sustainable solutions and continuous technological advancements. By acknowledging and engaging with the intricate phenomena of soil-structure interaction, while incessantly pursuing innovative and sustainable methodologies, we can tackle diverse challenges in foundation engineering. In doing so, we contribute substantially to the sustainable growth and resilience of our built environment.

10.4 Coupled numerical modeling

The intricate nature of soil-structure interaction necessitates the utilization of progressively sophisticated numerical modeling techniques. Looking ahead, we can anticipate further advancements in synergistic modeling approaches that account for the multi-physical and multi-phase behaviors of soils under diverse loading conditions [61].

10.5 Climate change and soil-structure interaction

The impacts of climate change, including sea-level rise and an increased incidence of extreme weather events, pose substantial challenges for soil-structure interaction. Future research endeavors will necessitate a concentrated effort to comprehend these effects and formulate resilient and adaptive foundation systems to mitigate potential harm [62].

10.6 Education and training in foundation engineering

Finally, the role of education and training in foundation engineering cannot be overstated. Future research should explore the development of innovative educational strategies, incorporating digital technologies into the process of teaching and learning foundation engineering principles [63, 64]. By adopting these research directions, we can aspire to enrich our understanding of soil-structure interaction phenomena, creating more efficient and sustainable solutions in foundation engineering. However, it's imperative to remember that these technological and scientific advancements should always be matched with a deep respect for nature and a firm commitment to social justice and equality.

11. Conclusion

Soil-Structure Interaction (SSI) continually presents substantial challenges to geotechnical engineers and scientists. As explored in this chapter, the nature of the issue is multifaceted, requiring interdisciplinary approaches and a profound understanding of the underlying soil mechanics and structural behavior. SSI extends beyond the application of intricate numerical models; it necessitates an in-depth comprehension of the fundamentals of geotechnical and structural engineering. Recent strides in computational methods, geotechnical investigation techniques, and an improved understanding of soil behavior under various loading conditions have greatly amplified our ability to analyze and design structures considering SSI effects. However, challenges persist, particularly in the realms of complex soil behavior, climate change effects, and sustainable practices in foundation engineering.

In an epoch of climate change and escalating infrastructure demands, it's essential to reconsider traditional methods and embrace innovative and sustainable solutions. Future research should concentrate on a more comprehensive understanding of SSI, incorporating considerations for environmental and social impacts. While new technological advancements offer considerable potential, their application should coincide with a profound understanding of their limitations and a dedication to ethical and sustainable practices. As we continue to erect and develop infrastructure globally, the understanding and application of soil-structure interaction will become progressively critical. By accepting the challenges and opportunities presented by this complex field, engineers can aid in sculpting a more sustainable and resilient future. Despite substantial progress in understanding and addressing the challenges associated with soil-structure interaction, there exist numerous opportunities for future research:

- *Holistic framework for SSI*: The need for a more comprehensive and integrated framework for understanding and analyzing SSI is clear. This framework should consider not only the geotechnical and structural aspects but also the environmental, social, and economic impacts of engineering decisions.

- *Climate change and SSI*: As the effects of climate change become increasingly discernible, there's a need to understand how shifts in weather patterns and sea levels will impact soil-structure interaction. Research in this area could center on understanding the impacts of the increasing frequency and intensity of extreme weather events on SSI and developing adaptive design strategies.

- *Sustainable geotechnical solutions*: In alignment with global sustainable development goals, there's a need for research into innovative and sustainable geotechnical solutions. This could include the usage of recycled or locally sourced materials in foundation construction or the evolution of low-carbon geotechnical engineering practices.

- *Advanced computational models for SSI*: While substantial advancements have been made in computational methods for analyzing SSI, challenges linger. Future research could concentrate on developing more advanced and reliable computational models that accurately capture the complex behavior of soils under varying loading and environmental conditions.

- *Integration of advanced technologies*: The integration of advanced technologies, such as Artificial Intelligence (AI) and Machine Learning (ML), into geotechnical engineering practice could revolutionize the way we understand and analyze SSI. These technologies bear the potential to provide valuable insights into complex soil behavior and enhance the accuracy and efficiency of SSI analysis.

By focusing on these research directions, we can further our understanding of soil-structure interaction and contribute to the development of safer, more resilient, and sustainable built environments.

12. Closing remarks

This chapter has delved into an in-depth examination of the complexities and challenges inherent in soil-structure interaction (SSI), a fundamental facet of geotechnical engineering. From deciphering basic principles and terminologies to dissecting advanced computational models and sustainable practices, this chapter endeavored to offer a thorough overview of the subject matter. The understanding and practical application of SSI are paramount to the safe and efficient design and construction of structures. As we persist in developing and building our infrastructure, the significance of considering SSI cannot be stressed enough. Neglecting the interactions between soil and structure can result in structural failures, over-design, or even under-design. Hence, it's vital to incorporate the effects of SSI into the design and analysis process. However, analyzing SSI is no simple task given the inherent complexities involved. A multitude of factors, including the type of soil, the properties of the structure, the nature of loads, and the environmental conditions, all play significant roles in determining the SSI. Thus, it's imperative for engineers and researchers to possess a comprehensive understanding of these factors and their impacts on SSI.

In an era characterized by climate change and mounting environmental concerns, sustainable practices in foundation engineering have become increasingly crucial. This chapter showcased several strategies and solutions that can be employed to make foundation engineering more sustainable. These encompass the use of recycled or locally sourced materials, the adoption of low-carbon practices, and the incorporation of resilience into design and construction. As we progress, it is expected that our understanding of SSI will continue to evolve in line with technological advancements, computational models, and sustainable practices. With further research and a steadfast commitment to sustainable and ethical practices, we can anticipate substantial progress in the field of foundation engineering.

Acknowledgements

We wish to extend our heartfelt gratitude to all the researchers, scientists, and engineers whose relentless dedication and diligence have driven progress in the field of soil-structure interaction (SSI). Their curiosity and unwavering commitment have cleared the path towards a more profound understanding of the intricacies of SSI, the creation of more sustainable practices in foundation engineering, and the design and construction of safer, more efficient structures. We are particularly grateful to the authors of all the studies, papers, and articles referenced in this chapter. Their invaluable contributions formed the fundamental basis for this comprehensive exploration of SSI. Lastly, we want to express our appreciation to our peers and colleagues for their valuable feedback and guidance throughout the writing process of this chapter. Their insights and recommendations significantly augmented the quality and breadth of our work.

Author details

Ali Akbar Firoozi* and Ali Asghar Firoozi
Department of Civil Engineering, Faculty of Engineering and Technology, University of Botswana, Gaborone, Botswana

*Address all correspondence to: a.firoozi@gmail.com

References

[1] Das BM, Sivakugan N. Principles of Foundation Engineering. 9th ed. Stamford, CT USA: Cengage Learning; 2018. ISBN-13: 978-1337705028

[2] Rad MM, Ibrahim SK. Optimal plastic analysis and design of pile foundations under reliable conditions. Periodica Polytechnica Civil Engineering. 2021;**65**(3):761-767. DOI: 10.3311/PPci.17402

[3] Sitharam TG, Kolathayar S, Ravi S, Matsagar V. Theory and Practice in Earthquake Engineering and Technology. Springer Tracts in Civil Engineering (SPRTRCIENG); 2022. DOI: 10.1007/978-981-19-2324-1

[4] Große C. A review of the foundations of systems, infrastructure and governance. Safety Science. 2023;**160**:106060. DOI: 10.1016/j.ssci.2023.106060

[5] Sun J, Zhu W, Mu B, Zhong J, Lin N, Chen S, et al. Efficient extraction of biodiesel feedstock and dehydration of kitchen waste: A method based on co-dissolution of liquefied dimethyl ether and water. Waste Management. 2022;**147**:22-29. DOI: 10.1016/j.wasman.2022.04.044

[6] Wang F, Wang Q, Adams CA, Sun Y, Zhang S. Effects of microplastics on soil properties: Current knowledge and future perspectives. Journal of Hazardous Materials. 2022;**424**:127531. DOI: 10.1016/j.jhazmat.2021.127531

[7] Salgado R, Kumar S, Basu D, Alvarez GME, Gavin K, Ahmad SF. General Report—Session 1: Analysis, Design, Testing and Performance of Foundations. 2008. Available from: https://scholarsmine.mst.edu/icchge/6icchge/session00c/1?utm_source=scholarsmine.mst.edu%2Ficchge%2F6icchge%2Fsession00c%2F1&utm_medium=PDF&utm_campaign=PDFCoverPages

[8] Kramer SL. Geotechnical earthquake engineering. Pearson Education India; 1996

[9] Wong HL. Dynamic soil-structure interaction. 1975. Available from: https://resolver.caltech.edu/CaltechEERL:1975.EERL-75-01

[10] Stewart JP, Fenves GL, Seed RB. Seismic soil-structure interaction in buildings. I: Analytical methods. Journal of geotechnical and geoenvironmental engineering. 1999;**125**(1):26-37. DOI: 10.1061/(ASCE)1090-0241(1999)125:1(26)

[11] Gazetas G. Analysis of machine foundation vibrations: state of the art. International Journal of soil dynamics and earthquake engineering. 1983;**2**(1):2-42. DOI: 10.1016/0261-7277(83)90025-6

[12] Vaseghiamiri S, Mahsuli M, Ghannad MA, Zareian F. Probabilistic approach to account for soil-structure interaction in seismic design of building structures. Journal of Structural Engineering. 2020;**146**(9):04020184. DOI: 10.1061/(ASCE)ST.1943-541X.0002741

[13] Falcone R, Lima C, Martinelli E. Soft computing techniques in structural and earthquake engineering: A literature review. Engineering Structures. 2020;**207**:110269. DOI: 10.1016/j.engstruct.2020.110269

[14] Younesian D, Hosseinkhani A, Askari H, Esmailzadeh E. Elastic and viscoelastic foundations: A review on linear and nonlinear vibration modeling and applications. Nonlinear Dynamics. 2019;**97**(1):853-895. DOI: 10.1007/s11071-019-04977-9

[15] Briaud JL. Geotechnical engineering: unsaturated and saturated soils. Hoboken, NJ: Wiley; 2013

[16] Zienkiewicz OC, Emson C, Bettess P. A novel boundary infinite element. International Journal for Numerical Methods in Engineering. 1983;**19**(3):393-404. DOI: 10.1002/nme.1620190307

[17] Beer J, van Geel B. Holocene climate change and the evidence for solar and other forcings. Natural Climate Variability and Global Warming: A Holocene Perspective. 2008;**138-162**. DOI: 10.1002/9781444300932

[18] Kausel E. Early history of soil–structure interaction. Soil Dynamics and Earthquake Engineering. 2010;**30**(9): 822-832. DOI: 10.1016/j. soildyn.2009.11.001

[19] Bolisetti C, Whittaker AS. Numerical investigations of structure-soil-structure interaction in buildings. Engineering Structures. 2020;**215**:110709. DOI: 10.1016/j.engstruct.2020.110709

[20] Shamsi M, Shabani MJ, Vakili AH. Three-dimensional seismic nonlinear analysis of topography–structure–soil–structure interaction for buildings near slopes. International Journal of Geomechanics. 2022;**22**(3):04021295. DOI: 10.1061/(ASCE)GM.1943-5622.0002301

[21] Mylonakis G, Gazetas G. Seismic soil-structure interaction: beneficial or detrimental? Journal of earthquake engineering. 2000;**4**(3):277-301. DOI: 10.1080/13632460009350372

[22] Shiravand MR, Parvanehro P. Spatial variation of seismic ground motion effects on nonlinear responses of cable stayed bridges considering different soil types. Soil Dynamics and Earthquake Engineering. 2019;**119**:104-117. DOI: 10.1016/j.soildyn.2019.01.002

[23] Firoozi AA, Naji M, Firoozi AA. Effects of soil–structure interaction on performance of bridges during earthquakes. Case Study: Integral Abutment Bridge in Pennsylvania, USA. Iranian Journal of Science and Technology, Transactions of Civil Engineering. 2023;**2023**:1-19. DOI: 10.1007/s40996-023-01122-w

[24] Jong SC, Ong DEL, Oh E. State-of-the-art review of geotechnical-driven artificial intelligence techniques in underground soil-structure interaction. Tunnelling and Underground Space Technology. 2021;**113**:103946. DOI: 10.1016/j.tust.2021.103946

[25] Reddy JN. Introduction to the Finite Element Method. 4th ed. Chicago, IL, USA: McGraw-Hill Education; 2019. ISBN-13: 978-1259861901

[26] Colaço A, Barbosa D, Costa PA. Hybrid soil-structure interaction approach for the assessment of vibrations in buildings due to railway traffic. Transportation Geotechnics. 2022;**32**:100691. DOI: 10.1016/j. trgeo.2021.100691

[27] Hashash YM, Park D. Non-linear one-dimensional seismic ground motion propagation in the Mississippi embayment. Engineering Geology. 2001; **62**(1–3):185-206. DOI: 10.1016/S0013-7952(01)00061-8

[28] Oliveto ND. Geometrically nonlinear analysis of friction pendulum systems under tri-directional excitation. Engineering Structures. 2022;**269**: 114770. DOI: 10.1016/j. engstruct.2022.114770

[29] Çerçevik AE, Avşar Ö, Hasançebi O. Optimum design of seismic isolation

systems using metaheuristic search methods. Soil Dynamics and Earthquake Engineering. 2020;**131**:106012. DOI: 10.1016/j.soildyn.2019.106012

[30] Loveridge F, McCartney JS, Narsilio GA, Sanchez M. Energy geostructures: A review of analysis approaches, in situ testing and model scale experiments. Geomechanics for Energy and the Environment. 2020;**22**:100173. DOI: 10.1016/j.gete.2019.100173

[31] Das BM. Principles of Geotechnical Engineering. 10th ed. Stamford, CT, USA: Cengage Learning; 2021. ISBN-13: 978-0357420485

[32] Ge B, Ruan H, Shu S, Zhang F, Gao Y. Effects of seismic amplification on the stability design of geosynthetic-reinforced soil walls. Geotextiles and Geomembranes. 2022;**50**(5):881-895. DOI: 10.1016/j.geotexmem.2022.05.003

[33] Freddi F, Galasso C, Cremen G, Dallsta A, Di Sarno L, Giaralis A, et al. Innovations in earthquake risk reduction for resilience: Recent advances and challenges. International Journal of Disaster Risk Reduction. 2021;**60**: 102267. DOI: 10.1016/j.ijdrr.2021.102267

[34] Sani AK, Singh RM, Amis T, Cavarretta I. A review on the performance of geothermal energy pile foundation, its design process and applications. Renewable and Sustainable Energy Reviews. 2019;**106**:54-78. DOI: 10.1016/j.rser.2019.02.008

[35] Chen Y, Guéguen P, Chen KH, Lin CJ, Ku CS, Huang WG, et al. Dynamic characteristics of TAIPEI 101 skyscraper from rotational and translation seismometers. Bulletin of the Seismological Society of America. 2023; **113**(2):690-709. DOI: 10.1785/0120220147

[36] Malhotra S. Seismic Soil-Pile Interaction: Physical Processes and Analytical Models. In: 5th International Conference on Recent Advances in Geotechnical Earthquake Engineering and Soil Dynamics. Missouri University of Science and Technology; 2010. p. 47. https://scholarsmine.mst.edu/icrageesd/05icrageesd/session05/47

[37] Coleman JL, Bolisetti C, Whittaker AS. Time-domain soil-structure interaction analysis of nuclear facilities. Nuclear Engineering and Design. 2016; **298**:264-270. DOI: 10.1016/j.nucengdes.2015.08.015

[38] Ahn S, Park G, Yoon H, Han JH, Jung J. Evaluation of soil–structure interaction in structure models via shaking table test. Sustainability. 2021;**13**(9):4995. DOI: 10.3390/su13094995

[39] Badr M, Lotfy A. Sustainable Tunneling and Underground Use. Cham: Springer Nature; 2019. DOI: 10.1007/978-3-030-01884-9

[40] Ayala F, Sáez E, Magna-Verdugo C. Computational modelling of dynamic soil-structure interaction in shear wall buildings with basements in medium stiffness sandy soils using a subdomain spectral element approach calibrated by micro-vibrations. Engineering Structures. 2022;**252**:113668. DOI: 10.1016/j.engstruct.2021.113668

[41] Ghosh B, Fatahi B, Khabbaz H, Nguyen HH, Kelly R. Field study and numerical modelling for a road embankment built on soft soil improved with concrete injected columns and geosynthetics reinforced platform. Geotextiles and Geomembranes. 2021;**49**(3):804-824. DOI: 10.1016/j.geotexmem.2020.12.010

[42] Ullah A, Kassim A, Ahmed S, Ullah R, Junaid M, Zia MD. Experimental

analysis of embankment on soft soil improved with bottom ash columns. Transportation Geotechnics. 2022;**37**: 100881. DOI: 10.1016/j.trgeo.2022.100881

[43] Naji M, Firoozi AA, Firoozi AA. A review: Study of integral abutment bridge with consideration of soil-structure interaction. Latin American Journal of Solids and Structures. 2020a; **17**:e252. DOI: 10.1590/1679-78255869

[44] Stefanidou SP, Sextos AG, Kotsoglou AN, Lesgidis N, Kappos AJ. Soil-structure interaction effects in analysis of seismic fragility of bridges using an intensity-based ground motion selection procedure. Engineering Structures. 2017; **151**:366-380. DOI: 10.1016/j.engstruct.2017.08.033

[45] Chen R, Congress SSC, Cai G, Duan W, Liu S. Sustainable utilization of biomass waste-rice husk ash as a new solidified material of soil in geotechnical engineering: A review. Construction and Building Materials. 2021;**292**:123219. DOI: 10.1016/j.conbuildmat.2021.123219

[46] Li Y, Sun H, Li D, Song J, Ding R. Effects of digital technology adoption on sustainability performance in construction projects: The mediating role of stakeholder collaboration. Journal of Management in Engineering. 2022;**38**(3): 04022016. DOI: 10.1061/(ASCE)ME.1943-5479.0001040

[47] Naji M, Firoozi AA, et al. "Soil arching" for piled embankments: Insights from stress redistribution behaviour of DEM modelling. Acta Geotechnica. 2020b;**15**:2117-2136. DOI: 10.1007/s11440-019-00902-x

[48] Brunelli A et al. Numerical simulation of the seismic response and soil–structure interaction for a monitored masonry school building damaged by the 2016 Central Italy earthquake. Bulletin of Earthquake Engineering. 2021;**19**(2):1181-1211. DOI: 10.1007/s10518-020-00980-3

[49] Van Ballegooy S, Malan P, Lacrosse V, Jacka ME, Cubrinovski M, Bray JD, et al. Assessment of liquefaction-induced land damage for residential Christchurch. Earthquake Spectra. 2014; **30**(1):31-55. DOI: 10.1193/031813EQS070M

[50] Al-Atroush ME, Sebaey TA. Stabilization of expansive soil using hydrophobic polyurethane foam: A review. Transportation Geotechnics. 2021;**27**:100494. DOI: 10.1016/j.trgeo.2020.100494

[51] Saif A, Ahmed Sheikh K. Design of grout curtain for temporary embankments on rock foundation. Geomechanics and Geoengineering. 2022;**17**(2):426-441. DOI: 10.1080/17486025.2020.1778193

[52] Pacheco J, de Brito J. Recycled aggregates produced from construction and demolition waste for structural concrete: Constituents, properties and production. Materials. 2021;**14**(19):5748. DOI: 10.3390/ma14195748

[53] Parlato M, Porto SM, Cascone G. Raw earth-based building materials: An investigation on mechanical properties of Floridia soil-based adobes. Journal of Agricultural Engineering. 2021;**52**(2):1-6. DOI: 10.4081/jae.2021.1154

[54] Han C, Yu XB. Analyses of the thermo-hydro-mechanical responses of energy pile subjected to non-isothermal heat exchange condition. Renewable Energy. 2020;**157**:150-163. DOI: 10.1016/j.renene.2020.04.118

[55] Tang C, Phoon KK. Statistics of model factors and consideration in

reliability-based design of axially loaded helical piles. Journal of Geotechnical and Geoenvironmental Engineering. 2018; **144**(8):04018050. DOI: 10.1061/(ASCE) GT.1943-5606.0001894

[56] Zhang Y, Hu X, Wang Y, Jiang N. A critical review of biomineralization in environmental geotechnics: Applications, trends, and perspectives. Biogeotechnics. 2023;**2023**:100003. DOI: 10.1016/j.bgtech.2023.100003

[57] Agunwamba JC, Onyia ME, Nwonu DC. Development of expansive soil geopolymer binders for use in waste containment facility. Innovative Infrastructure Solutions. 2021;**6**:1-18. DOI: 10.1007/s41062-020-00400-0

[58] Guo D, Zhu X, Xie J, Zhang C, Zhao Z, Zheng L. Undergraduate program for urban underground space engineering in China: Exploration and practice. Tunnelling and Underground Space Technology. 2021;**116**:104084. DOI: 10.1016/j.tust.2021.104084

[59] Dang LC, Khabbaz H, Ni BJ. Improving engineering characteristics of expansive soils using industry waste as a sustainable application for reuse of bagasse ash. Transportation Geotechnics. 2021;**31**:100637. DOI: 10.1016/j. trgeo.2021.100637

[60] Liu L, Cai G, Zhang J, Liu X, Liu K. Evaluation of engineering properties and environmental effect of recycled waste tire-sand/soil in geotechnical engineering: A compressive review. Renewable and Sustainable Energy Reviews. 2020;**126**:109831. DOI: 10.1016/j.rser.2020.109831

[61] Cheng F, Sun X, Wu P, Chen Z, Yu T, Liu W, et al. A fully coupled thermo-hydro-mechanical-chemical model for methane hydrate bearing sediments considering the effect of ice. Journal of Marine Science and Engineering. 2023;**11** (4):766. DOI: 10.3390/jmse11040766

[62] Pan Y, Zhang L. Roles of artificial intelligence in construction engineering and management: A critical review and future trends. Automation in Construction. 2021;**122**:103517. DOI: 10.1016/j.autcon.2020.103517

[63] Sigahi TF, Sznelwar LI. From isolated actions to systemic transformations: Exploring innovative initiatives on engineering education for sustainable development in Brazil. Journal of Cleaner Production. 2023;**384**: 135659. DOI: 10.1016/j. jclepro.2022.135659

[64] Naji M, Firoozi AA. Performance of bridge envelope during earthquake using finite element and artificial neural network techniques. The Open Civil Engineering Journal. 2022;**16**(1):1-26. DOI: 10.2174/18741495-v16-e2208100

Chapter 5

Geosynthetic Reinforcement Applications

Erol Guler

Abstract

Geosynthetic applications range from geotechnical to geo-environmental areas and include countless cases. In this paper, the aim was to introduce the reader to the reinforcement applications. A few examples of reinforcement were chosen to demonstrate the frequently used applications and to draw the attention of the reader to the vast variety of other reinforcement applications. First, a brief introduction to geosynthetics is given. As an example of a very frequently applied reinforcement application, information is given on geosynthetic reinforced soil retaining walls. Especially the benefits of geosynthetic-reinforced retaining walls under earthquake loading conditions are emphasized. Then, some information is given on using geosynthetic reinforcement below foundations to improve the bearing capacity. As a third and final example, the use of geosynthetic reinforcement as an encasement around stone columns is mentioned. The improvement of the geosynthetic encased columns compared to ordinary stone columns under earthquake loading conditions was shown.

Keywords: geosynthetic, reinforced soil, geotextile, geogrid, retaining wall, embankment, geosynthetic encasement

1. Introduction

Geosynthetic is a generic term describing a product, at least one of whose components is made from a synthetic or natural polymer, in the form of a sheet, a strip, or a three-dimensional structure, used in contact with soil and/or other materials in geotechnical and civil engineering applications [1]. The concept of geosynthetics started with the introduction of geotextiles and has become a family that includes products, such as geogrid, geonet, geomembrane, geosynthetic clay liners (GCL), geofoam, geocell, prefabricated vertical drain (PVD), geotube, geosynthetic encased stone column, and geocomposites. The fact that the products are made of polymer material creates an important advantage because both the polymer properties and manufacturing procedures can be adapted to get the best properties for a required engineering solution. Especially in earthworks, polymers are very efficient, because soil protects them from UV light, which has a tendency to deteriorate the properties of the polymers. However, of course, they are also used successfully in applications open to the atmosphere under appropriate manufacturing and design conditions. Geosynthetic products are produced in a wide variety of forms and features and are produced in the most successful way to fulfill the intended function in the end use. Today, geosynthetics are used successfully in many civil engineering, geotechnical engineering, transportation engineering, environmental engineering, and hydraulic projects. Geosynthetics provide in many applications superior engineering solutions and at the same time reduce construction time and cost as well. Geosynthetics also make infrastructure more sustainable. They extend the service life of roads, reduce the use of aggregates, conserve and protect water, minimize land disturbance, and control soil erosion. Geosynthetics are used in a wide variety of applications, such as roads, railways, airports, embankments, retaining walls, reservoirs, canals, dams, erosion prevention projects, sediment control, solid and liquid waste landfill base and top liners, production facilities and tailing dams of mining activities, aquaculture, and agriculture.

Detailed descriptions of each of the various geosynthetic products mentioned above will not be provided within this article due to space limitations. However, the definitions of various geosynthetic products and applications can be found in Ref. [1]. Naturally, there are geosynthetics produced using a wide variety of production methods and materials within each main product group. For example, geotextiles can be classified as woven, nonwoven, or knitted. For woven geotextiles, there are various methods of weaving and also woven yarns can be for example in the form of filaments, tapes, etc. The nonwoven geotextiles can also be produced using different techniques like needle-punching or by thermal bonding, also called spun-bonding. In addition, the polymer used in the production of geosynthetics can also vary. Polypropylene and polyester are most commonly used in the production of geotextiles, but many different polymers can also be used. Again, for example, geogrids can be manufactured as woven, extruded, or welded, and can be made using a wide variety of polymers as needed. Naturally, although the production method and raw material remain the same, it is possible to produce products with varying properties. For example, tensile strength, permeability, etc., can vary significantly for the same production type and polymer. Likewise, there are various geomembranes produced with different raw materials and surface properties. This variability is a major advantage of geosynthetics. The scientific and technological level reached in polymer engineering today allows the production of polymer materials to be produced with the required features, so that the products can satisfy the needs of the civil engineer, who will design and manufacture facilities with these products. So, the combination of the proper raw material property and the production technique can help the geosynthetic to fulfill its required function in the most optimum way. While this wide variety of products is a very important advantage, it also creates a handicap, because civil engineers in general, but even geotechnical engineers, are not fully aware of the functions and design aspects of geosynthetic products, so it is difficult to determine what would be the most suitable geosynthetic for a project among products that are produced with a wide variety of raw materials and production techniques and have a wide range of mechanical and hydraulic properties. To overcome this complexity, it is important for geotechnical engineers to be well informed about geosynthetics. In fact, the International Geosynthetics Society IGS (International Geosynthetics Society) is a nongovernmental organization established just for this very purpose, namely to disseminate knowledge for the proper use of geosynthetics [2]. I would like to add that sustainability is at the heart of what IGS does. The publications, events, lectures and research highlight the many ways that geosynthetics contribute to the sustainable development goals. Although not included in this paper, because of space limitations, how the geosynthetics contribute to resolving the sustainability crises have been addressed in many papers and a synthesis paper that has been published recently [3].

In this paper only a sample of geosynthetic reinforcement applications will be highlighted to introduce the general reader about the versatile uses of geosynthetics.

2. Reinforced soil retaining structures

As it is known, retaining walls or slopes create level differences, which are often necessary for civil engineering projects. Historically, masonry walls were used as the only alternative. Later on, reinforced concrete retaining walls took over as the predominant retaining wall type. If there is enough space, an alternative is to fill or excavate with a slope. Today, it is a very common practice to keep excavations stable at steeper angles by equipping them with the so-called soil nails. In fill slopes, the most common method is to use geosynthetic reinforcement. The slope inclination can be varied infinitely with the help of the geosynthetic reinforcement up to geosynthetic-reinforced retaining walls with totally vertical facings (**Figure 1**). This brings flexibility to the design by optimizing the space available to reduce cost.

Design details of geosynthetic-reinforced walls are not mentioned in this article, because these are too comprehensive to fit into the scope of this article, and also there are many specifications giving the details of the design [4–6].

Although the widespread use of geosynthetic-reinforced soil retaining walls is due to economic incentives, it is now known that they also exhibit a much better engineering performance, especially in earthquake-prone regions, and therefore they are preferred over rigid reinforced concrete retaining walls [7]. A case study [8] states that a total of approximately 3500 m long geosynthetic reinforced soil wall has been manufactured alone on the Hokkaido Shinkansen high-speed train line in Japan, and the maximum height of these walls is 11 m. In the same project, 29 bridge abutments with a maximum height of 13.4 m were formed as geosynthetic-reinforced walls. An integrated bridge with geosynthetic reinforcement was also constructed (**Figure 2**). This type of geosynthetic-reinforced integrated bridges is becoming increasingly common. The American highway administration has also published a design specification for the design of geosynthetic reinforced bridge abutments, which is called Geosynthetic Reinforced Soil-Integrated Bridge System (GRS-IBS) [9].

It is a well-accepted fact that geosynthetic-reinforced soil retaining structures are also much less demanding in terms of foundation soil bearing capacity because they

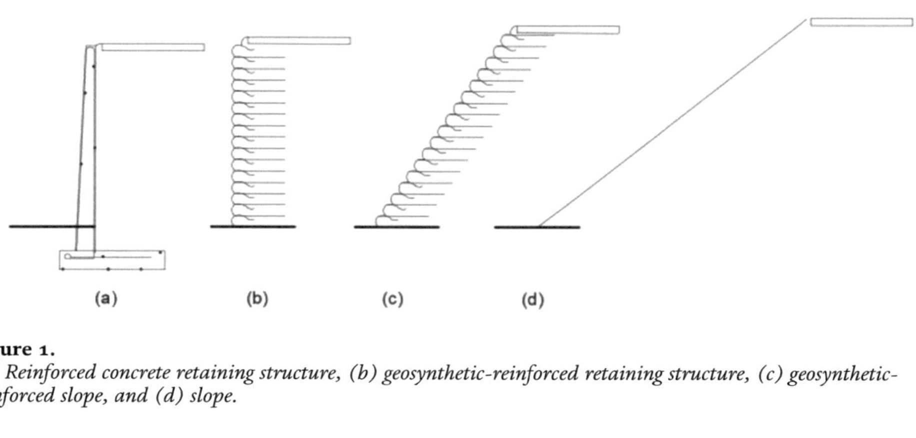

Figure 1.
(a) Reinforced concrete retaining structure, (b) geosynthetic-reinforced retaining structure, (c) geosynthetic-reinforced slope, and (d) slope.

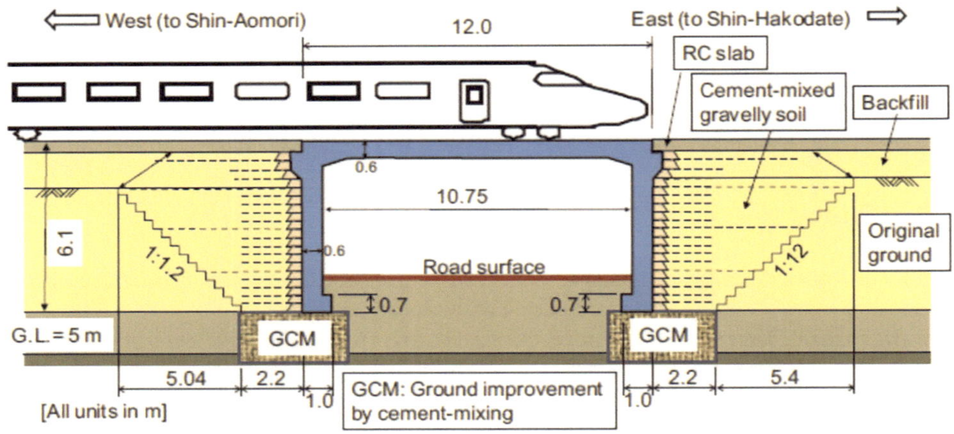

Figure 2.
Integrated bridge with geosynthetic reinforcement in Hokkaido Shinkansen line [8].

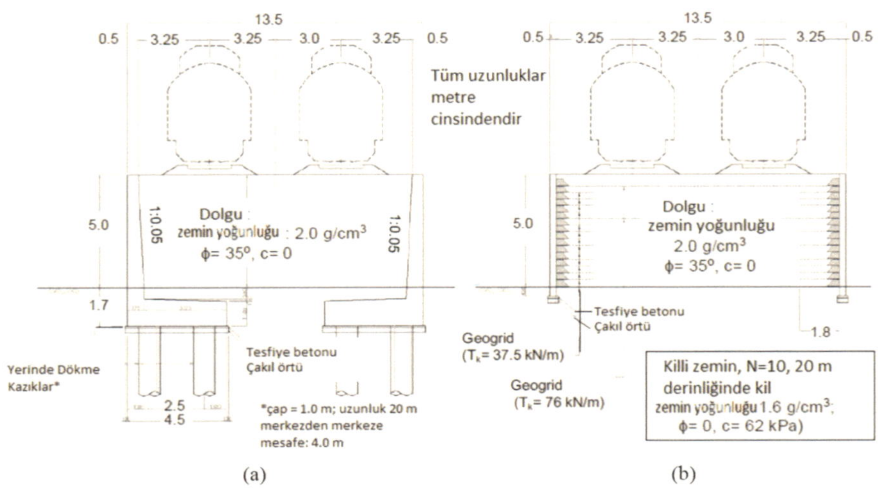

Figure 3.
Embankment wall alternative design comparison from Hokkaido Shinkansen line [8].

do not exert an overturning moment to the foundation soil, which tends to create the maximum stress at the toe of the retaining structure. To highlight this advantage, in the comparative example given in Ref. [8], it can be clearly seen that for the reinforced concrete retaining wall system shown in **Figure 3a**, a pile foundation was necessary since the foundation soil was quite weak. However, in the solution given in **Figure 3b**, namely when the retaining walls are designed as geosynthetic-reinforced walls, a surficial foundation was found to be sufficient. This essentially indicates that the geosynthetic-reinforced wall technology provides an advantage, especially in retaining walls that need to be built on soft soils. It goes without saying how much of an advantage it would be to save both time and cost to be spent on pile foundation construction.

The information briefly summarized above shows that geosynthetic-reinforced retaining walls offer solutions that provide great advantages in terms of both engineering performance and economy. Another advantage of geosynthetic-reinforced retaining walls is that they can easily adapt to a wide variety of geometric conditions. Since they are not built-in panels like reinforced concrete retaining walls, they can easily adapt to the steep slopes existing along the length of the wall at the foundation level (**Figure 4**), and the top elevations of the walls to the desired geometry.

As geosynthetic-reinforced retaining walls can easily adapt to geometric changes in the longitudinal direction, they can as easily adapt to geometric constraints on the cross-section as well. An interesting example to demonstrate this flexibility in cross-section is seen in **Figure 5**. In a reinforced concrete wall, if there is a need for stairs connecting the bottom and top elevations, the only alternative is to build a separate reinforced concrete ladder structure. However, as can be seen in **Figure 5**, the staircase has been created within the wall by systematically shifting the facing out of the wall at different elevations.

Because in reinforced concrete systems, it is not possible to place the foundation on the filled ground, the stem has to cover the elevation difference as one piece. However, as we can see in the example in **Figure 3**, retaining walls with geosynthetic reinforcement can easily be built on the embankment since they will exert nearly

Figure 4.
Example of a geosynthetic-reinforced retaining wall constructed in a V-shaped valley.

Figure 5.
Geosynthetic reinforced soil retaining structures with built-in stairs.

uniform vertical pressure on the foundation soil. In the figure, the distance between the two facings can vary between zero and any value (**Figure 6**). Undoubtedly, necessary analyses should be performed depending on this distance [4].

Finite element analyses were used to investigate several parameters for a parametric study [10]. The results of the finite element analyses indicated that, under working loads, the maximum shear stresses occur along the classical expected potential failure plane. However, when in the finite element analysis the shear strength is gradually reduced up to failure (phi-c reduction), no internal stability problem within the reinforced soil zone is observed. The classical failure plane expected to occur in the unreinforced backfill, indicating an external stability problem (**Figure 7**).

As it is known, in practice often a suitable soil or rock excavated locally is used as a backfill in geosynthetic-reinforced retaining walls. Hence, it is important to understand how the cohesive backfill will affect the behavior of geosynthetic-reinforced walls. In this respect, Ref. [11] reports furthermore the behavior of geosynthetic-reinforced walls with clayey and granular backfills under earthquake loads. One of the most important results of the study is that if cohesive soil is used instead of granular soil as backfill, wall deformations are reduced by nearly 50%. This finding indicates that cohesive soils can be a suitable backfill material if proper drainage and

Figure 6.
Example of a tiered wall on a bridge approach fill. All the stair structure has also been constructed using the reinforced soil system.

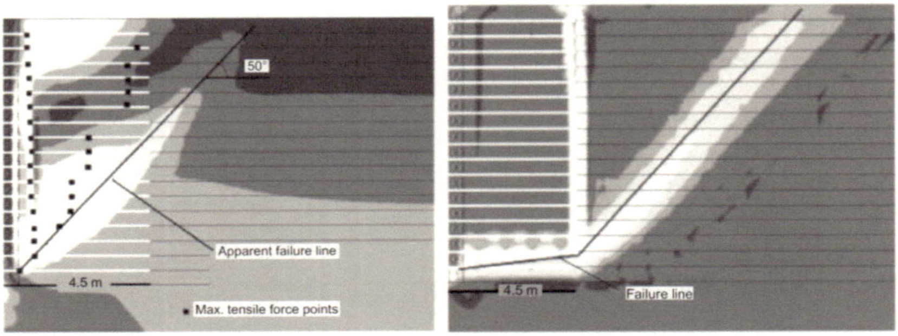

(a) (b)

Figure 7.
Incremental shear strain contours: (a) at end of construction, and (b) at end of ϕ–c reduction in the same model [10].

compaction are provided. The parametric study further showed that lower displacements were obtained for higher-strength reinforcements, but the tensile forces on the reinforcements increased. Another finding mentioned by the paper is that doubling the earthquake acceleration magnitude caused the lateral displacements to increase more than twice, indicating that the acceleration and displacements do not have a linear relationship (**Figure 8**).

This fact is also supported by the numerical study, where it is reported that less tensile forces occur in reinforcements in cohesive backfills than in granular backfill. As a result of the numerical model study, it has been shown that cohesive soils can also be a suitable filling material if the appropriate drainage measures are taken [11].

Ref. [12] tested 1.9 m high, ½ scaled reinforced soil walls with concrete modular block-facing elements on a shaking table. The models were exposed to the scaled El-Centro earthquake record. Two different reinforcement lengths, L/H = 0.6 and L/H = 0.9, were used in the experiments. As a result, it was concluded that the geosynthetic-reinforced walls performed well under earthquake loading and no permanent displacement was observed at the end of the seismic loading. The acceleration measurements indicated that for all models the accelerations at the facing increased with increasing height of the wall. Measured accelerations at different time intervals are seen in **Figure 9**. It can be seen that the accelerations vary at different times, but the general trend that accelerations increase with increasing height is seen clearly.

Ref. [12] further showed that the failure plane is in good agreement with the Rankine failure plane as can be seen in **Figure 10**.

Ref. [13] tested eight modular concrete blocks under earthquake loads again on a shaking table. A total of eight models were tested: four 2 meters tall and ½ scale, one 1-meter tall and ¼ scale and three 2 meters tall and ¼ scale. All the walls had concrete modular blocks as facing elements. Experimental results showed that reinforcement length and density did not affect the amplification factors much. Face displacements are also less affected as long as the reinforcement length meets the minimum requirements in the FHWA design criteria. No significant permanent displacement was observed at the end of the experiments. For short reinforcements, the measured

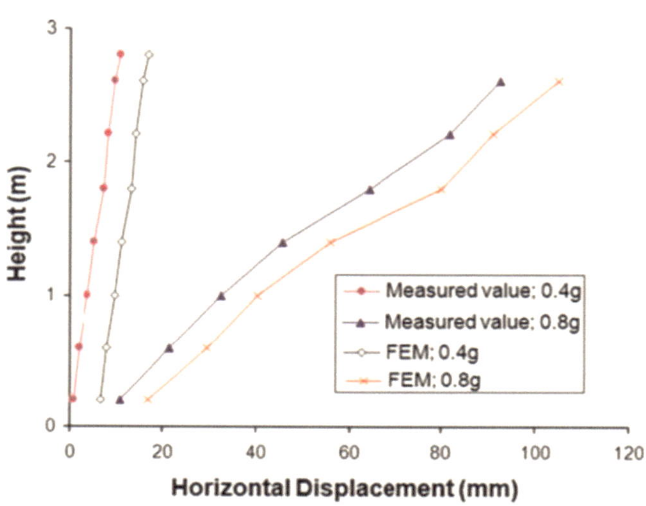

Figure 8.
Horizontal displacements for accelerations of 0.4 and 0.8 g [10].

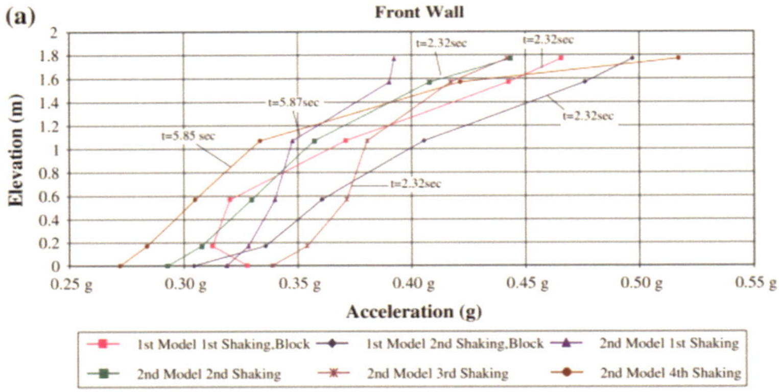

Figure 9.
Maximum accelerations measured on the front wall of various walls shaken by the El-Centro earthquake [12].

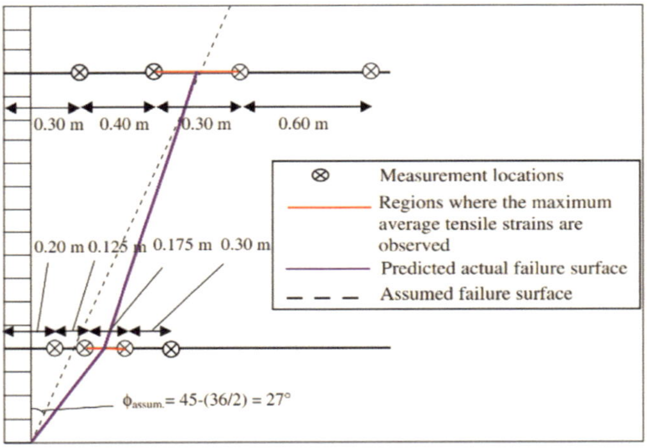

Figure 10.
Internal failure surface determination and the regions where peak average tensile strains are observed during the El-Centro excitations [12].

tensile stress on reinforcements was higher than the tensile stresses calculated in the design. At the end of the study, the failure surfaces obtained by combining the locations of the maximum stresses on the geotextiles and the failure planes suggested by the Rankine theory were compared (**Figure 11**). It can be seen that the failure planes were in general in good agreement with the Rankine plane. The only exception was seen for the walls with longer reinforcements, where the potential failure plane developed closer to the facing.

In another series of shaking table tests, the model of a 3.8 m high prototype wall from a real design has been tested [14]. Because of the payload capacity of the shaking table, a scaling factor of 2 is used. So, four ½ scale and 1.9 meters high reinforced soil retaining wall models were tested under 1-g loading conditions. Their aim was to investigate the effects of various parameters on the seismic performance of the wall. The parameters investigated were the magnitude of horizontal acceleration, reinforcement length and strength, type of backfill soil (sand versus gravel), and face inclination (vertical versus 6-degree batter). The model walls were instrumented as

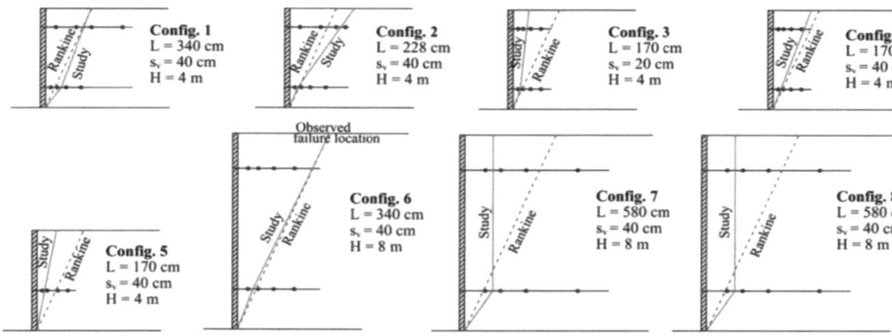

Figure 11.
Potential failure surfaces determined from geotextile strains compared with the Rankine failure plane shown on the prototype walls represented by the tested model configurations (values beside illustrations are given for the prototype wall) [13].

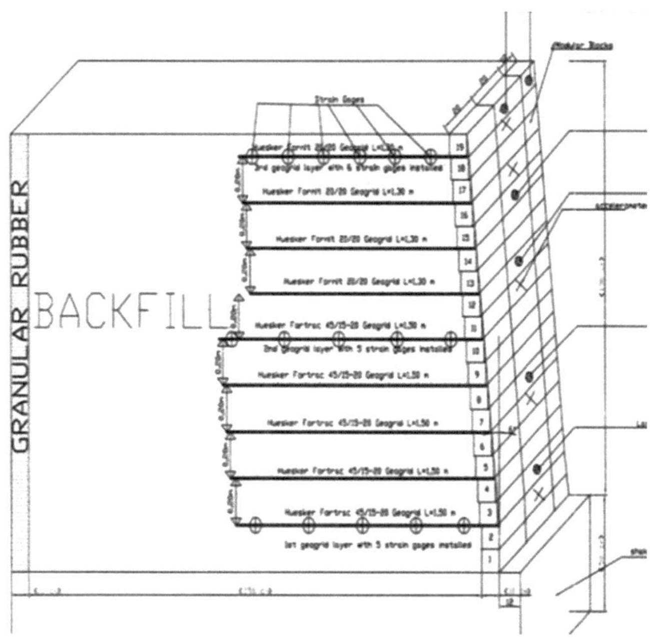

Figure 12.
Schematic of the tested model wall (Guler and Basbug, 2012) [15].

follows: i) nine accelerometers, five of which were installed on the facing elements of the wall and three on top of the backfill and one accelerometer on the shaking table, ii) six laser displacement sensors to measure the displacement of the wall face and, iii) sixteen strain gages installed on three different geogrid reinforcement layers (bottom layer, at mid-height layer, and top layer). A schematic view of the model is seen in **Figure 12**. The model was shaken by scaled versions of El-Centro earthquake. When the amplitude was increased to 250% of the original record, the accelerations reached almost a level of 1 g. Even under these extreme accelerations, the walls manufactured in accordance with the current regulations showed a successful performance. Even at very extreme horizontal acceleration values, no stability problem occurred, and

	Gravel backfill		Sand backfill	
	Peak displacement (mm)	**Residual displacement (mm)**	**Peak displacement (mm)**	**Residual displacement (mm)**
100% El-Centro	1.9	0.5	2.2	0.6
250% El-Centro	28.1	4.7	30.6	4.3

Table 1.
Peak and permanent displacements under various earthquake magnitudes [15].

Figure 13.
Field monitored GRS-IBS bridge abutments built in Virginia [16].

the deformations remained within acceptable limits. The maximum and residual (permanent) displacements at these loading values are given in **Table 1**.

Ref. [16] studied the components of the Geosynthetic Reinforced Soil-Integrated Bridge Abutment (GRS-IBS) method suggested by Ref. [9] with a finite element analysis. For the finite element model, a fully instrumented real prototype bridge abutment was used to calibrate and validate the finite element model (**Figure 13**).

Effects of the presence of a bearing bed and reinforced soil foundation (RSF), the vertical reinforcement spacing and reinforcement stiffness, and subgrade compressibility on design components, including the lateral displacement of facing, maximum tension in reinforcement (T_{max}), connection strength (T_o), and differential settlement in reinforced soil structures was investigated [16]. It was determined that the inclusion of a bearing bed, which is a densely reinforced zone below the superstructure, contributes to the reduction in the lateral displacement of facing. The bearing bed also reduces the maximum tension in reinforcements by 20%. The presence of the bearing bed decreased the settlement difference between the superstructure and integrated approach zones for the reinforcement spacing for the wall, which had a reinforcement spacing of S_v = 0.4 m. The vertical reinforcement spacing is an important variable that controls the behavior of the geosynthetic-reinforced soil walls. Decreasing S_v from 0.4 m to 0.2 m decreases the lateral displacement of the facing by up to 1.8 times, the maximum tension in the reinforcement by up to 2.2 times, and the connection strength by up to 4 times. It was also determined that the magnitude of the pressure applied from the superstructure affects the distribution of T_{max} with depth. When the

GRS was modeled with S_v = 0.2 m and nonuniform reinforcement distribution (a typical GRS-IBS structure), the results showed that the distribution of T_{max} under typical working load conditions (surcharge of approximately 40 kPa) shows a uniform trend with depth but with a high bridge superstructure load (approximately 200 kPa) the trend shifts to become a bilinear distribution. However, when the GRS was modeled with Sv = 0.4 m and uniform reinforcement distribution (a typical MSE structure) the distribution, regardless of the magnitude of the superstructure loads, shows a linear distribution with depth. This behavior difference can be most likely attributed to the stiffer behavior due to closely spaced reinforcement. **Figure 14** shows changes in lateral displacement of facing with depth. A comparison of the results from the model with reinforced soil foundation (RSF) and ordinary foundation (NRSF) shows that the presence of RSF reduces the lateral displacement of the facing by up to 60%. It was further recommended based on the parametric study that the connection strength must be considered in design even in closely spaced reinforced soil structures. As summary, the inclusion of the bearing bed and reinforced soil foundation was found to have beneficial effects in reducing the lateral displacement of the facing, the maximum tension in the reinforcement, the connection strength, the foundation settlement, and the settlement difference within the abutment.

As mentioned above, in many projects, cohesive soils or weathered rock is used as a backfill because it is locally available and therefore a great saving both in terms of money, and probably more important in terms of CO2 footprint can be achieved. Having seen that the cohesive soil can be efficiently implemented as a backfill in geosynthetic reinforced walls, [17] investigates the seismic performance of geosynthetic-reinforced modular block retaining walls backfilled with cohesive, fine-grained clay-sand soil mixture on a shaking table. Tests were performed for three ½ scaled (wall height 1.90 m) and ¼ scaled model walls to investigate the effects of backfill type, the influence of reinforcement length, and reinforcement stiffness effects. The El-Centro and Kobe earthquake records of varying amplitudes were used

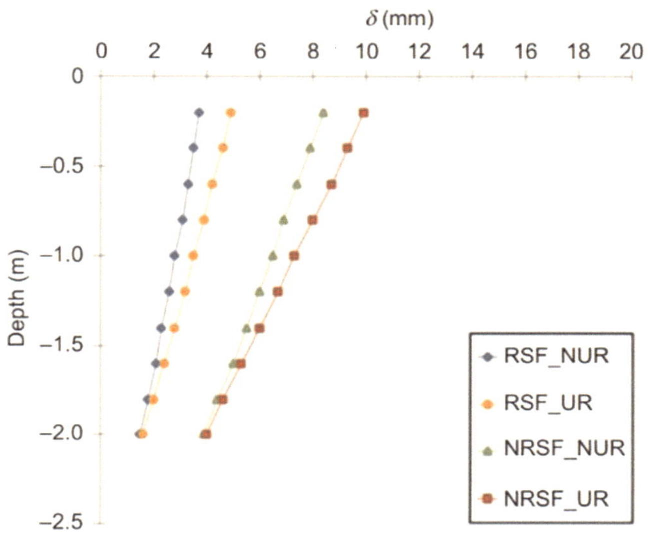

Figure 14.
Effect of foundation soil reinforcement suggested by GRS-IBS on horizontal deformation of the wall [16].

as base acceleration. In conclusion, when relative displacement values were compared, model walls with cohesive backfill exhibited superior seismic performance compared to the model walls with granular backfill. As seen in **Table 2**, the maximum displacements observed even under 125% Kobe earthquake were lower than the maximum deformation, which occurred under 100% Kobe earthquake in sand backfill. When the results of reinforcement stress measurements were compared, it was seen that seismic loading-induced strain demand on geogrids in cohesive backfill was as high as their counterparts embedded in granular backfill. No remarkable impact of reinforcement length was observed during the tests for cohesive backfills. For all tested model walls, no failure and extreme deformation were observed even under extreme seismic loads. The model walls exhibited minimal residual deformations on the facing as a result of the applied base excitations for both cohesive and granular backfill. Seismic loading-induced settlements were observed, however only at the back of the model. Some minimal tension cracks were detected in the unreinforced backfill zone for cohesive backfills. The seismic performance of all setups could be acknowledged as successful when the results were compared with the limit values reported in the literature. The authors provide the following important disclaimer: "It must be stated that these favorable results were observed in the protected laboratory environment, where atmospheric conditions are controlled. In order to transfer the advantage of cohesive backfill into practice, proper drainage measures must be taken to prevent the wetting of the cohesive backfill."

Although retaining structures are commonly analyzed in two dimensions because they are relatively long structures, in some cases, due to property boundaries, it may be necessary to include corner turns in geosynthetic-reinforced retaining walls. This corner turns can be induced at a right angle or greater/smaller angles. Ref. [18] reports about some cracking and separation of facing blocks in the near vicinity of a right-angle corner of a 13 m high real wall as seen in **Figure 15**.

In order to analyze this problem, a 3D finite element model was built using 467,221 elements. However, this model could not be run on a windows-based computer. So, a scaled-down model was constituted to understand the mechanisms involved (**Figure 16**). In conclusion, it was seen that the soil modulus and reinforcement stiffness both play important roles in the cracking of facing blocks.

As seen in **Figure 17**, as the modulus of the backfill soil increases, the deviator stress, which can be influential on the cracking of the facing elements, reduces significantly. Similarly, the relationship between reinforcement stiffness and deviator stress acting on the facing blocks as shown in **Figure 18** indicates that the reinforcement stiffness is equally effective in reducing the deviator stress acting on the facing block. So, it is recommended that in order to reduce the cracking of the facing blocks at corner turns, the backfill should be compacted as much as possible and a stiffer geosynthetic reinforcement should be used. Furthermore, the results of the 3D finite

Backfill Soil	L/H	Base excitation	Peak displacement (mm)
Sand	0.8	100% Kobe Earthquake	34.6
30% Clay and 70% sand	0.8	125% Kobe Earthquake	32.4
30% Clay and 70% sand	0.6	125% Kobe Earthquake	34.3

Table 2.
Properties of model walls and measured maximum displacements [17].

Figure 15.
Geosynthetic-reinforced soil wall with concrete block facing elements [18].

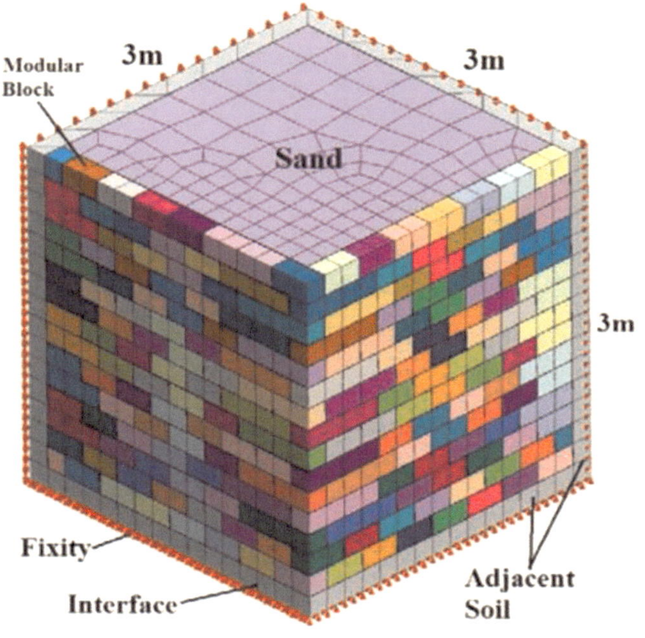

Figure 16.
3D finite element model of the corner of a geosynthetic-reinforced wall with block-facing elements [18].

element model indicated that the wall undergoes a curved form in the plan and this causes the separation of the blocks.

It was determined that the abrupt change in the slope, which is also responsible for the cracking of the blocks, occurs only when the reinforcement stiffness is relatively low. When the stiffness was increased, the abrupt change in the slope ratio suddenly

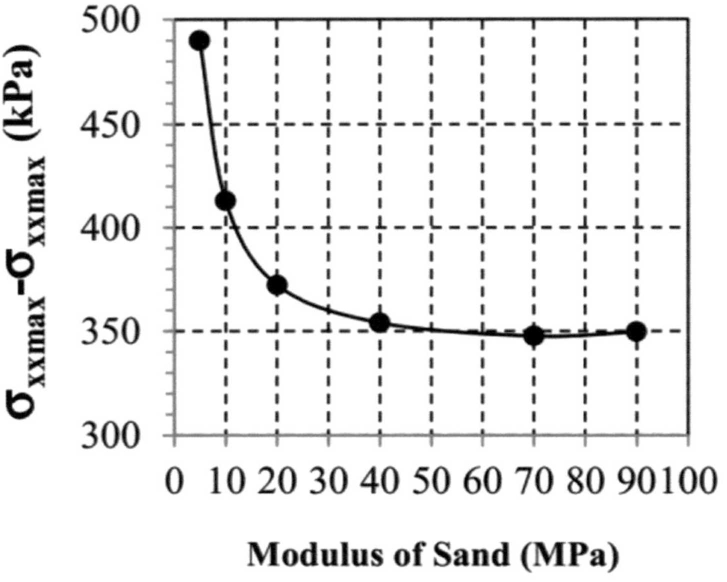

Figure 17.
Maximum deviator stress acting on facing blocks versus soil modulus [18].

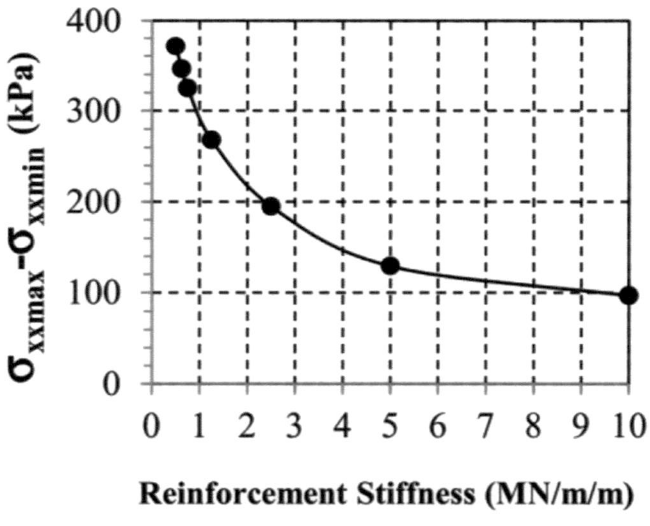

Figure 18.
Maximum deviator stress acting on facing blocks versus reinforcement stiffness [18].

diminished (**Figure 19**). This indicates that there should be a minimum reinforcement stiffness that is required for the integrity of the blocks on the wall corner.

3. Reinforced embankments

Another reinforcement application of geosynthetics is reinforced foundations. This can be a very economical solution if the bearing capacity of the soil is not sufficient to

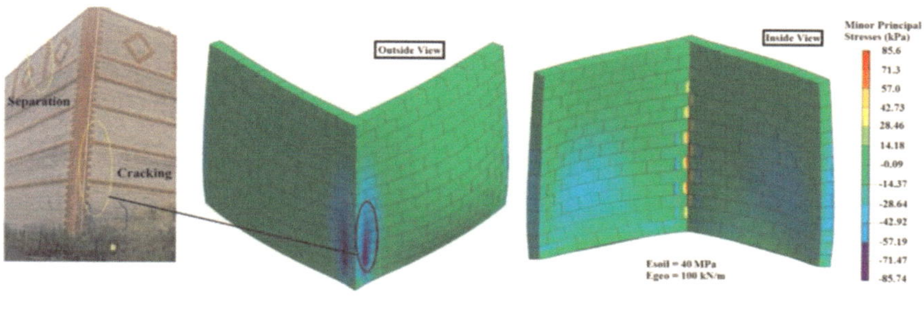

Figure 19.
Horizontal stress concentration was observed close to the corner in the FE model, where block cracking was observed in the real wall [18].

carry the foundation loads safely, because the other alternatives are improving the bearing capacity with solid inclusions, which are relatively costly compared to just using one or several layers of geosynthetic below the foundation. There are many scientific publications reporting on theoretical and experimental studies to understand the role of reinforcement materials in improving the bearing capacity of foundation soils. Experimental studies to evaluate the bearing capacity of footings on reinforced sandy soil were reported by Refs. [18–25]. These studies show that the reinforcement configuration values that give the maximum bearing capacity value depend on soil and footing types. Ref. [20] investigated the influence of the model footing diameter and embedment depth on the bearing capacity of circular shallow footings by centrifugal model testing. Ref. [19] performed laboratory tests to determine the behavior of geosynthetic-reinforced sandy soil foundations and studied the effect of different parameters contributing to their performance. The ultimate bearing capacity of a circular footing, placed over a soil mass reinforced with horizontal layers of circular reinforcement sheets, was determined with the limit analysis in conjunction with finite elements and linear optimization by Ref. [21]. The critical positions and corresponding optimum diameter of the reinforcements to achieve maximum bearing capacity were established, and a marked improvement in the bearing capacity is evident in the case of two layers of reinforcements rather than a single layer of reinforcement. All these studies show the importance of reinforcement for the bearing capacity of foundations. Several researchers have investigated the degree of improvement achieved with different reinforcement lengths. Ref. [22] determined the optimum length of the reinforcement layer as L = 2.5B for a square footing on geogrid-reinforced sand. Ref. [23] found the optimum reinforced length for strip footing for reinforced sand and reinforced clay to be L = 8B and L = 5B, respectively. Ref. [24] stated that the optimum reinforcement length is between L = 5B and L = 7B. Ref. [25] found the optimum reinforcement length to be L = 5B. As seen from the studies in the literature, there is no unique value proposed for the reinforcement length. If one was to accept the conclusions of the previous studies, the optimum length of the reinforcement layer for maximum bearing capacity lies anywhere between L = 2B and L = 8B.

Only a few studies have compared the behavior of different reinforcement types [22, 26, 27]. Thus, the effect of reinforcement type has not been investigated thoroughly. However, the performance of reinforced soil foundation depends on the interaction between the soil and the reinforcement and, therefore, the properties of the geogrid play a major role.

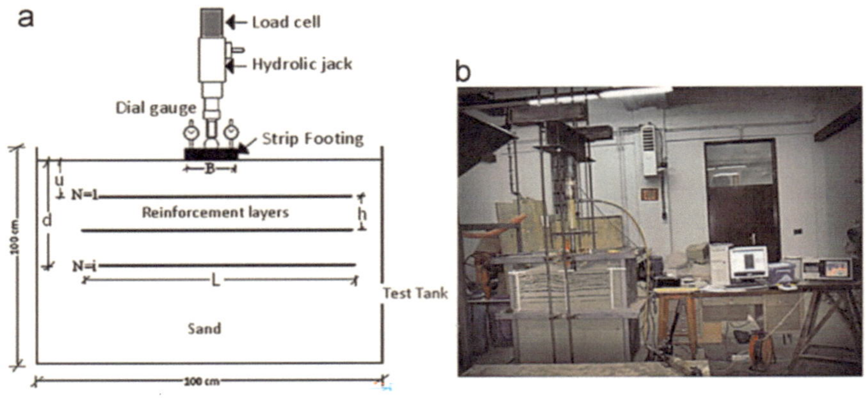

Figure 20.
Test setup (a) schematic diagram, and (b) photo of test setup [28].

Ref. [28] conducted laboratory model tests using four different geosynthetic reinforcements in order to understand the effect of reinforcement type along with the effects of the depth of the first reinforcement layer, the vertical spacing between consecutive layers of reinforcement, the total number of reinforcement layers, and the width of the geosynthetic reinforcement. The test setup is seen in **Figure 20**. The soil used in the tests was a dry poorly graded sand (C_u = 2.5) and was placed into the model box using a raining technique, at a relative density of 46%. As reinforcement, four different geosynthetic products were used. One of them was a polypropylene woven geotextile with an ultimate tensile strength of 60 kN/m. The remaining three were geogrids. Geogrid 1 and Geogrid 2 were woven polyester geogrids with an ultimate tensile capacity of 35 and 55 kN/m respectively. Geogrid 3 was an extruded polypropylene geogrid with an ultimate tensile capacity of 45 kN/m.

Ref. [28] concluded that the improvement in the bearing capacity due to reinforced foundation soil is typically reported for the ultimate condition. However, from a practical point of view, we have also to limit the amount of vertical settlement. As such, it is important to note the vertical settlement, which occurs with increases in the bearing capacity. The results of the test indicated that the amount of bearing capacity at small settlements and large settlements varies significantly and that one product showing a very good improvement under large settlements may give poor results under small settlements and *vice versa*.

Figure 21 shows that for all reinforcement types, for all number of reinforcement layers, the bearing capacity of the reinforced soil foundation increased with increasing settlement ratios. Especially at smaller settlement ratios (s/B = 0.05–0.2) and for one to four reinforcement layers, the woven geogrid gave a higher improvement in the bearing capacity compared to extruded geogrid. The Bearing Ratio values of geotextile reinforced soils were smaller than those of the other reinforced types at all settlement ratios. The authors further highlighted that as the number of reinforcement layers increased, the increment of the Bearing Ratio values rose and it had a linear effect. For N = 5, the maximum BR occurred for Geogrid 1 reinforced sand, but the minimum BR value occurred for Geogrid 3.

Based on the results of many experimental data, Ref. [29] suggested a limit equilibrium method to determine the bearing capacity of strip foundations on geosynthetic-reinforced sand soils. When a reinforcement of the foundation soil is

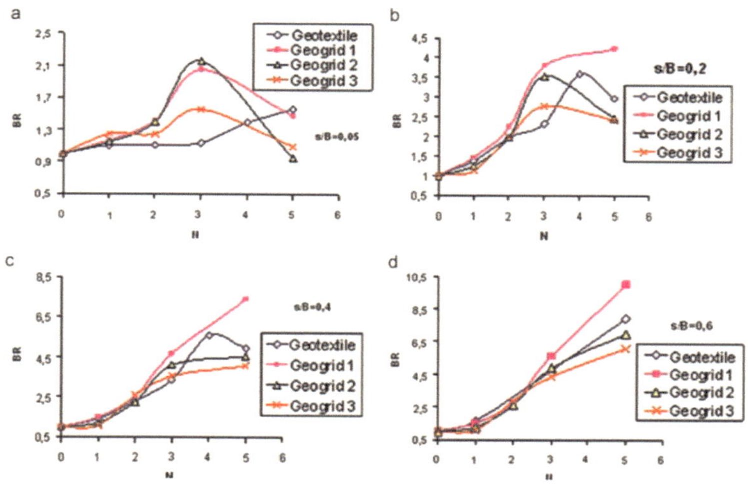

Figure 21.
Comparison of different reinforcement types to bearing ratio for reinforcement length of L/B = 3 and different settlement ratios (a) s/B = 0.05, (b) s/B = 0.2, (c) s/B = 0.4, and (d) s/B = 0.6 [28].

considered, the geosynthetic reinforcement will be embedded into the soil. Therefore, in general, there are two soil layers below the foundation, namely the natural soil and the fill soil. Therefore, the limit equilibrium approach that allows the calculation of the bearing capacity of a two-layered soil was adapted to analyze the reinforced foundation problem. Since the soil above the reinforcement will be compacted properly, it was considered that the natural soil is less dense than the fill soil. The geometry and failure concept adapted is shown in **Figure 22**.

Although the literature suggests that it is beneficial to extend the reinforcement outside the footprint of the foundation, construction restraints may not allow this and this will lead to the condition that the reinforcement can be placed only as wide as the foundation itself. Therefore, the formula developed also considers the reinforcement length, which is equal to the foundation width (L/B = 1). The most important unknown in the bearing capacity calculation is the activated reinforcement stress. Ref. [29] suggested a relation obtained by the multiple regression as shown in Eq. (1).

$$T = 110\ (h/B) + 68.8\ B + 5.07\ \phi + 8.05\ (L/B) - 267 \tag{1}$$

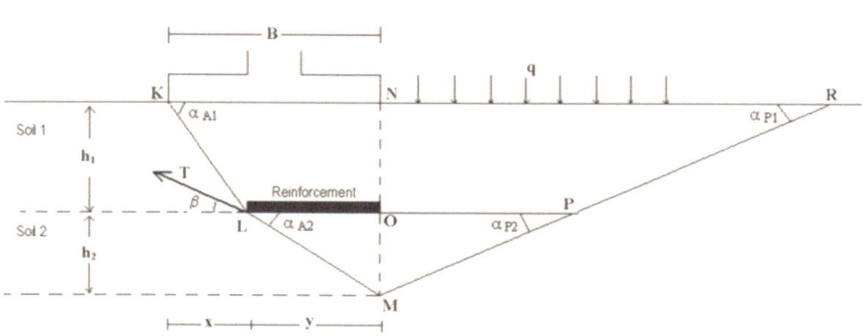

Figure 22.
Simplified slip surface with coulomb wedges to analyze a geosynthetic-reinforced foundation [29].

where T is the tensile stress in the reinforcement, h is the distance of the reinforcement from the base of footing, B is the footing width, ϕ is the internal friction angle of the upper strata, and L is the reinforcement length.

Along with the bearing capacity, the amount of allowable settlement is also very important in determining the allowable bearing capacity of soil. In this respect, the pressure distribution below a foundation resting on reinforced soil must be known. However, the stress distribution below foundations resting on reinforced soil foundations was investigated very little. In the literature related to reinforcing unpaved roads, the general assumption is that the reinforcement helps to distribute the load to a larger area and hence reduces the settlement. However, these studies are in general related to pavement analyses, and hence consider well-compacted granular materials, rather than loose sands, which necessitate a soil improvement. Consequently, the stress distribution in medium dense and loose sand is not considered. Ref. [30] provides important information on the stress distribution below the foundation based on experimental data and finite element verifications. Although stress values measured in the laboratory test are slightly smaller than FEM in some conditions, the authors conclude that there is a good correlation between the experimental measurements and the FE analysis results. The most important finding of this research was that the presence of reinforcements transfers vertical stress to deeper levels. However, it must be noted that a difference in the behavior exists depending on the reinforcement type and the reinforced zone depth.

4. Geosynthetic encased columns (GECs)

Ground improvement using ordinary stone columns (OSCs) in soft clayey soils is a cost and time-efficient soil remediation technique that has been practiced for decades. OSCs have been used in a wide spectrum of applications for enhancing the foundation soil properties of rigid and flexible structures, such as buildings, embankments, and oil storage tanks, that are founded on weak clays [31]. OSCs are also used efficiently used in mitigating the liquefaction risk of loose saturated sands [32]. While OSCs have proven themselves to be viable method, there are inherent shortcomings associated with the use of OSCs in soft soils. Early studies have pointed out that their stability is predominantly based on the available lateral support that is provided by the surrounding soil. When implemented in extremely soft soils (s_u < 15 kPa), the columns usually fail in bulging due to the lack of lateral support that the weak soil can offer. One way to overcome bulging failure is to encase the granular column materials with a reinforcing geosynthetic and thereby forming a geosynthetic encased column (GEC), which increases column performance by providing lateral confinement [33, 34]. The engineering behaviors of OSCs and GECs have been studied by means of both laboratory and field experiments, finite element methods, and analytical models. Most of the laboratory tests reported in the literature deal with the physical modeling of columns with small-scale models [35]. Also, field tests of encased columns have been conducted and it was demonstrated that the presence of encasement around the column enabled the column to support 2.3 times the total applied vertical stress [36].

Although there is plenty of literature on the vertical stress-strain behavior of GECs and OSCs, there are not many studies in the literature on the behavior of GECs and OSCs under the action of dynamic loads. Ref. [37] conducted a finite element analysis to model GECs under the action of seismic input motions and determined that implementation of GECs to support embankments underlain by weak soils greatly reduces the seismically induced settlements. The aim of this study is to shed light on the seismic behavior of GECs and OSCs supporting earthen embankments underlain by weak clay.

Because of the limitations of finite element analysis, it is not possible to simulate the intrusion of the stones of the stone column into the soft clay and thereby expand in volume and hence reduce in stiffness. Therefore, it is obvious that the improvements determined with the help of finite element analysis show only a partial contribution of the geosynthetic encapsulation limited to the increased stiffness due to the confining it applies to the stone column. However, with the use of three-dimensional finite element analysis, it has been determined that the total settlement under earthquake loading conditions can be reduced with the help of geosynthetic encapsulation. It was also shown that the GECs undergo less lateral displacements than the ordinary stone columns. Finite element analyses also show that significant tensile stress develops in the geosynthetic encapsulation, indicating the prevention of further bulging under earthquake loading conditions. In summary, the encapsulation of stone columns by a geotextile reinforcement (instead of installing ordinary stone columns), helps with the integrity of the column in very soft soils and reduces the system displacements under seismic impact.

Ref. [38] investigated the seismic behavior of GECs and OSCs installed in soft clay under embankments. The clay had an undrained cohesion of 4.25 kN/m^2 and the geosynthetic encasements GT1, GT2, and GT3 had a secant modulus of 18, 440, and 1150 kN/m, respectively at a strain of 10%. 1-g shaking table tests are conducted on OSCs and GECs. **Figure 23** shows the results from stress-controlled vertical load tests of all columns and the clay beds after the seismic excitations phase. The difference in pressure-settlement behavior of ordinary columns and columns reinforced with GT1 is not significant. This is because GT1 offered very little confinement therefore including GT1 as a media to attach the strain rosettes did not introduce significant error to the assumption that columns reinforced with GT1 in fact behave like ordinary

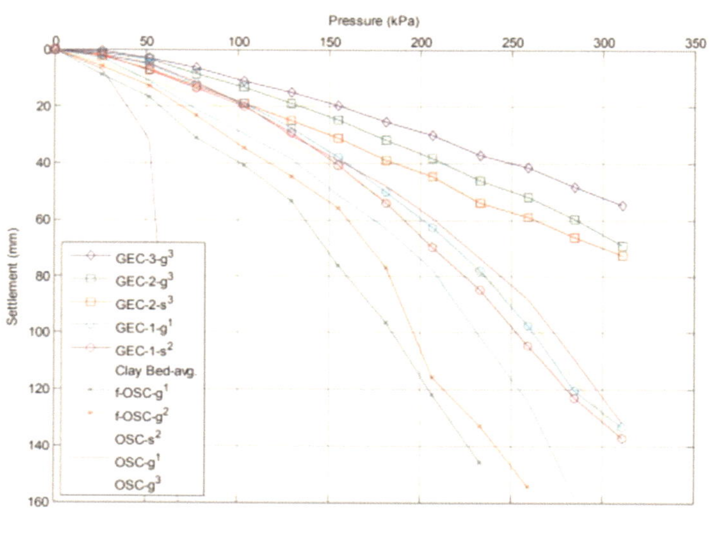

Figure 23.
Load settlement curves of clay, OSC, and GECs after the system has been exposed to earthquake loading conditions [38].

stone columns. GEC-3 with gravel infill experienced a minimum settlement value of about 55 mm at the end of the stress-controlled loading phase. GEC-2 experienced larger maximum settlements of 73 and 68 mm, for sand and gravel infill, respectively.

For the testing of inclusions into soft clay under earthquake loading conditions, the boundaries of the model must satisfy the free field condition. Therefore, flexible containers or more specifically laminar boxes constitute the vast majority of the cases reported in the literature for models tested with shaking tables. Such laminar boxes are able to efficiently model soil's free field response to input motions. The laminar box reported in Ref. [39] is square in plan with inner clearances of 900 $\times$ 900 mm. The height of the samples that can be accommodated in the laminar box is 1932 mm. A 300 mm deep rigid base cavity underlay the laminates (**Figure 24**). The laminar box consists of 16 individually supported laminates, which are made up of aluminum sigma profiles. A vertical clearance of 2 mm is provided between the laminates. The sigma profiles used are rectangular in cross-section with a width and height of 50 and 100 mm, respectively. The laminar box design makes use of the hollow space at the center of the sigma profiles and smooth guide rods are placed throughout the sigma profiles in the direction of shaking. Teflon riders are fitted in the ends of the aluminum profiles so that there is minimum resistance to horizontal movement. A laminar box commissioned as such, enables purely horizontal laminate movements without causing any rocking or twisting with respect to the vertical axis and diminishes the possibility of cantilever deformations and the toppling of the laminar box. Since the sigma profiles have hollow cross-sections, a significant reduction in mass is achieved while retaining the flexural rigidity of the laminates, which provides unyielding boundaries for the housed soil specimen. The resulting laminates have an assembly-to-soil mass ratio of about 7.5%, which ensures that the inertia effects of the laminates are within tolerable limits to study the 1D response of the soil.

When an impact-type input acceleration is applied to the table, it reaches a peak value at each laminate at a given time. Using this information, it was possible to determine the delay of peak value for all laminate levels. This information made it possible to determine the speed of the upward traveling shear wave or, in other words, the shear wave velocity.

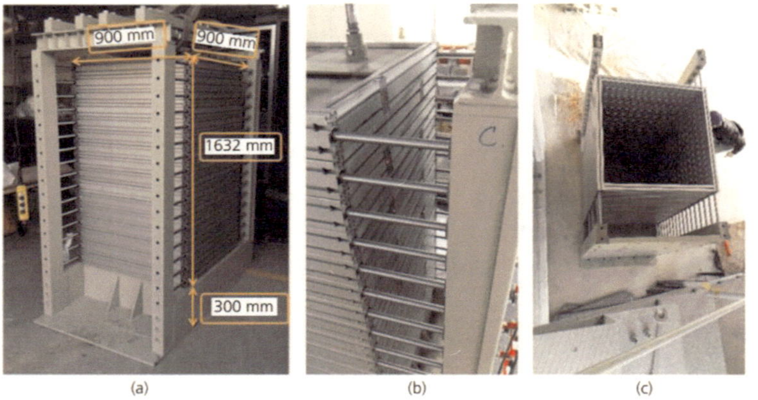

Figure 24.
(a) Dimensions of the laminar box, (b) close-up image of the guide rods and laminates, and (c) top view of the laminar box [39].

Using the measured shear wave velocities, the small strain shear modulus (G_{max}) of the soft soil composite inside the laminar box could be calculated by making use of Eq. (2).

$$Gmax = \rho {V_s}^2 \tag{2}$$

where ρ is the mass density of soil and V_s is the shear wave velocity. Using an average bulk mass density of 1630 kg/m^3 for all models tested, the small strain shear modulus for models of clay, clay with OSC, and clay with two different geosynthetic encasements were calculated to be 115, 259, 595, and 978 kPa, respectively. These results indicate that the inclusion of high modulus GECs at an area replacement ratio of 7% increased the small-stiffness shear modulus of a soft clay bed by almost an order of magnitude.

The shear modulus of the clay and the OSC or GEC-enhanced clay were also determined by plotting the shear strain and shear stress loops given in **Figure 25**. **Figure 25** illustrates the hysteresis loops extracted for the 10th cycle of 0.5 Hz sinusoidal excitation. The secant shear stiffnesses values for models with clay, clay improved with OSC and clay improved with a less stiff encasement and clay improved with a stiffer encasement are 14.5, 21.1, 39.6, and 90.6 kPa, respectively.

Most of the available experimental data pertain to the engineering behavior of GECs under vertical loading as briefly summarized above. However, especially the rigid inclusions, such as piles, OSCs, or GECs, close to the edge experience shear stresses. This is a well-known fact and studied for all embankment designs under the concept of edge failure. Although this constitutes a real threat to the stability of both OSCs and GECs, there are only limited studies on the behavior of such columns under shear loading [40].

Ref. [41] developed a unit cell shear device to study the behavior of OSCs and GECs behavior under shear forces. This device has a maximum height of 1850 mm and a diameter of 460 mm. The device allows for the installation of a single granular column with a diameter of 113 mm at the center of the unit cell. This corresponds to an area replacement ratio (the ratio of column area to unit cell area) of 6%, which is a lower-bound value in field applications. The device can be operated to shear the specimen under static loads or under cyclic loads. In the tests along with the pure loose sand also soil improved with OSCs and GECs were tested. For GECs, three types of geosynthetics have been used, namely with secant modulus values of 36, 380, and

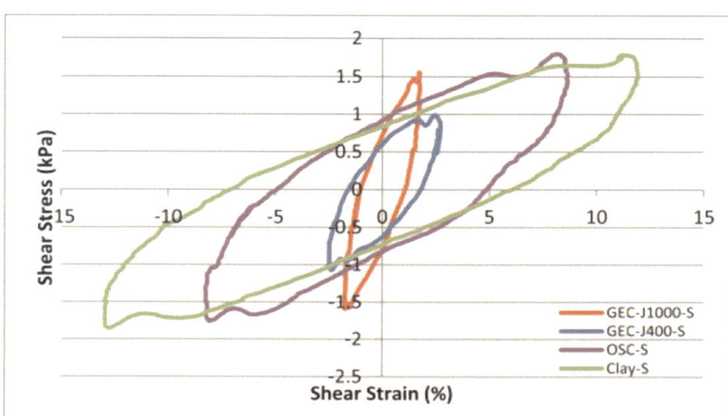

Figure 25.
Hysteresis loops for models consisting of clay alone and clay bed with OSC and GECs [39].

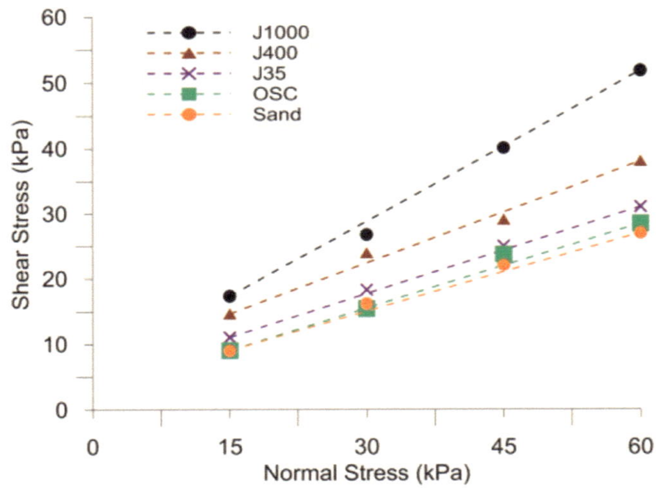

Figure 26.
Mohr-coulomb failure envelopes for the static shear tests [41].

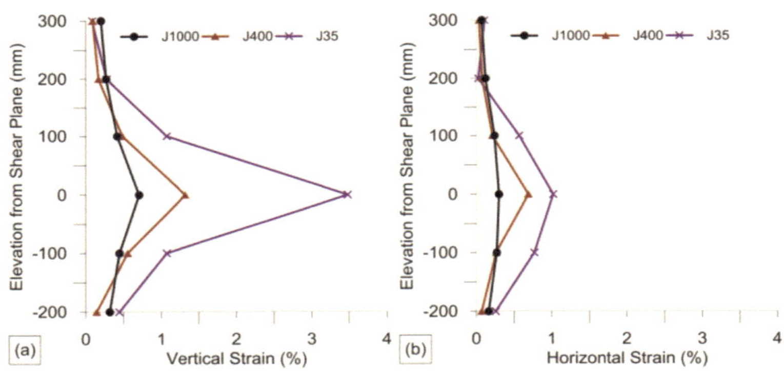

Figure 27.
Maximum values of (a) vertical and (b) horizontal strains were measured in static shearing of unit cells over the height of the columns [41].

1050 kN/m at 2% strain and are denoted as J35, J400, and J1000, respectively. The Mohr-Coulomb failure envelopes obtained are shown in **Figure 26**. From **Figure 26**, it is clear that the shear resistance of stone columns is significantly improved with the help of the encasement. It can be seen, that the internal friction angle of the loose sand (23.3°) could only be increased to 24.7° with stone columns alone. However, the overall internal friction angle for the unit cell was increased to 28° and 35° for geosynthetic encasements of J400 and J1000, respectively. The cohesion intercepts varied between 2 and 8 kPa.

The encasements of OSCs were equipped with strain gages to determine how the shear plane affects the columns. From **Figure 27** it can be seen that the shear applied on a single plane affects the column along a height approximately four times the diameter of the OSCs and GECs. It is also seen from the same figure that the GEC encasements are both strained in the vertical and horizontal directions. Especially the significant strain in the vertical direction hints at the tensile stress developed, which shows that the tensile strength in the encasement significantly increases the shear resistance of the column.

5. Conclusion

Reinforcement applications using geosynthetics cover a very wide area. In this synthesis paper, it was tried to highlight some of the more common and frequent reinforcement applications as well as some less frequently used reinforcement applications.

In general, the use of geosynthetics allows the engineer to come up with an alternative design, which is generally more economical, takes a shorter time to construct, and reduces the CO2 footprint.

The reader is encouraged to learn more about the advantages geosynthetics provide to geotechnical construction to be able to optimize their design.

Obviously, being a construction material, the properties of each geosynthetic product should be well understood and demonstrated with help of tests. Furthermore, appropriate QA/QC procedures have to be followed during the installation.

Last but not least, since geosynthetics allows the engineer to come up with innovative solutions, it also provides joy and professional satisfaction to the engineer.

Author details

Erol Guler
George Mason University, Fairfax, VA, USA

*Address all correspondence to: fguler@gmu.edu

References

[1] ISO ISO 10318-1 Geosynthetics— Tertms and Definitions. International Organization for Standardization. 2015

[2] Available from: https://www.geosyntheticssociety.org/

[3] Touze N. Healing the world: A geosynthetics solution. Geosynthetics International. 2021;**28**(1):1-31

[4] Berg R, Christopher B, Samtani N. Design of Mechanically Stabilized Earth Walls and Reinforced Soil Slopes. Washington DC: FHWA NHI-09-083; 2009

[5] BS 8006-1 Code of Practice for strengtened/reinforced soils and other fills, Bitish Standards, 2010

[6] German Geotechnical Society. Recommendations for Design and Analyses of Earth Structures Using Geosynthetic Reinforcement (EBGEO). Essen: Ernst & Sohn GmbH; 2011

[7] Koseki J, Bathurst R, Guler E, Kuwano J, Maugeri M. Seismic stability of reinforced soil walls. In: 8th International Conference on Geosynthetics. Japan: Yokohama; 2006

[8] Tatsuoka F. Design, construction and performance of GRS structures for railways in Japan. In: Invited Lecture, First National Conference on Geosynthetics. Honduras: The Honduran Chapter of IGS; 2014

[9] Adams M, Nicks J, Stabile T, Wu T, Schlatter W, Hartmann J. Geosynthetic Reinforced Soil Integrated Bridge System, Interim Implementation Guide, FHWA-HRT-11-026. McLean, Virginia, USA: US Department of Transportation, Federal Highway Administration; 2011

[10] Güler E, Hamderi M, Demirkan M. Numerical analysis of reinforced soil-retaining wall structure with cohesive and granular backfills. Geosynthetics International. 2007;**14**(6):330-345

[11] Guler E, Cicek E, Demirkan M, Hamderi M. Numerical analysis of reinforced soil walls with granular and cohesive backfills under cyclic loads. Bulletin of Earthquake Engineering. 2012;**10**(3):793-811

[12] Guler E, Enunlu A. Investigation of dynamic behavior of Geosynthetic reinforced soil retaining structures under earthquake loads. Bulletin of Earthquake Engineering. 2009;**7**:737-777

[13] Guler E, Selek O. Reduced-scale shaking table Tets on Geosynthetic reinforced soil walls with modular facing. Journal of Geotechnical Geoenvironmental Engineering. American Society of Civil Engineers (ASCE). 2014;**140**(6):64

[14] Basbug E, Cengiz C, Guler E. 1-g shaking table tests to determine the behavior of geosynthetic reinforced soil walls under seismic loads. Transportation Geotechnics. 2021;**30**

[15] Guler E, Alexiew D, Basbug E. Dynamic behavior of geogrid reinforced segmental block walls under earthquake loads. In: 5th International Conference on Earthquake Geotechnical Engineering. Chile: Santiago; 2011

[16] Gebremariam F, Tanyu B, Guler E, Urgessa G, Shen P. Numerical investigation of reinforced soil structures. Geosynthetics International. 2021;**28**(1):95-112

[17] Kilic I, Cengiz C, Edincliler A, Guler E. Seismic behavior of Geosynthetic-

reinforced retaining walls backfilled with cohesive soil. Geotextiles and Geomembranes. 2021;**49**(5):1256-1269

[18] Hamderi M, Guler E, Raouf A. An investigation on the formation of cracks at the corner turns of the modular block earth walls. International Journal of Civil Engineering. 2019;**17**:219-230

[19] Abu-Farsakh M, Chen Q, Sharma R. An experimental evaluation of the behaviour of footings on geosynthetic-reinforced sand. Soils & Foundations. 2013;**53**(2):335-348

[20] Toyosawa Y, Itoh K, Kikkawa N, Yang J, Liu F. Influence of model footing diameter and embedded depth on particle size effect in centrifugal bearing capacity tests. Soils & Foundations. 2013;**53**(2):349-356

[21] Chakraborty M, Kumar J. Bearing capacity of circular foundations reinforced with geogrid sheets. Soils & Foundations. 2014;**54**(4):820-832

[22] Guido V, Chang D, Sweeney M. Comparison of geogrid and geotextile reinforced earth slabs. Canadian Geotechnical Journal. 1986;**23**(4): 435-440

[23] Das B, Shin E, Omar M. The bearing capacity of surface strip foundations on geogrid reinforced sand and clay—A comparative study. Geotechnical and Geological Engineering. 1994;**12**(1):1-14

[24] Ghosh A, Ghosh A, Bear A. Bearing capacity of square footing on pond ash reinfored with jute-geotextile. Geotextiles and Geomembranes. 2005; **23**(2):144-173

[25] Dawson A, Moghaddas Tafreshi S. Comparison of bearing capacity of a strip footing on sand with geocell and with planar forms of geotextile reinforcement. Geotextiles and Geomembranes. 2010;**28**:72-84

[26] Chen Q. An Experimental Study on Characteristics and Behavior of Reinforced Soil Foundation, Ph.D. Dissertation. Baton Rouge, Louisiana, USA: Louisiana State University; 2007

[27] Latha M, Somwanshi A. Effect of reinforcement form on the bearing capacity of square footings on sand. Geotextiles and Geomembranes. 2009; **27**(6):409-422

[28] Cicek E, Guler E, Yetimoglu T. Effect of reinforcement length for different geosynthetic reinforcements on strip footing on sand soil. Soils and Foundations. 2015;**55**(4):661-677

[29] Cicek E, Guler E. Bearing capacity of strip footing on reinforced layered granular soils. Journal of Civil Engineering and Management. 2015; **21**(5):605-614

[30] Cicek E, Guler E, Yetimoglu T. Stress distribution below a continuous footing on geotextile-reinforced soil. ASCE International Journal of Geomechanics. 2018;**18**(3)

[31] Manohar R, Patel S. Ground improvement with stone columns—A review. In: Advances in Civil Engineering. Lecture Notes in Civil Engineering. Vol. 83. Springer: Singapore; 2021. pp. 197-217

[32] Adalier K, Elgamal A, Meneses J, Baez J. Stone columns as liquefaction countermeasure in non-plastic silty soils. Soil Dynamics and Earthquake Engineering. 2003;**23**(7):571-584

[33] Raithel M, Kirchner A, Schade C, Leusink E. Foundation of constructions on very soft soils with geotextile encased

columns-state of the art. In: Geo-Frontiers. Texas, USA: Austin; 2005

[34] Alexiew D, Brokemper D, Lothspeich S. Geotextile encased columns (GEC): Load capacity, geotextile selection and pre-design graphs. In: Geo-Frontiers. Texas, USA: Austin; 2005

[35] Miranda M, Da Costa A. Laboratory analysis of encased stone columns. Geotextiles and Geomembranes. 2016; **44**(3):269-277

[36] Hosseinpour I, Almeida M, Riccio M. Full-scale load test and finite-element analysis of soft ground improved by geotextile-encased granular columns. Geosynthetics International. 2015;**22**(6): 428-438

[37] Guler E, Alexiew D, Abbaspour A, Koc M. Seismic performance of Geosynthetic-encased stone columns. Transportation Research Record. 2014; **2462**:77-88

[38] Cengiz C, Guler E. Seismic behavior of geosynthetic encased columns and ordinary stone columns. Geotextiles and Geomembranes. 2018;**46**(1):40-51

[39] Cengiz C, Guler E. Shaking table tests on geosynthetic encased columns in soft clay. Geotextiles and Geomembranes. 2018;**46**(6):784-758

[40] Murugesan S, Rajagopal K. Shear load tests on stone columns with and without geosynthetic encasement. Geotechnical Testing Journal. 2009; **32**(1)

[41] Cengiz C, Kilic I, Guler E. On the shear failure mode of granular column embedded unit cells subjected to static and cyclic shear loads. Geotextiles and Geomembranes. 2019;**47**(2):193-202